# Biotechnology and its Application in Agricultural Science

## ABOUT THE BOOK

Agricultural biotechnology is a collection of scientific techniques, including genetic engineering, that are used to create, improve, or modify plants, animals and microorganisms. Agriculture refers to the production of food and goods through farming and forestry. Agriculture was the key development that led to the rise of civilization, with the husbandry of domesticated animals and plants creating food surpluses that enabled the development of more densely populated and stratified societies. The study of agriculture is known as agricultural science. Agricultural scientists are incorporating biotechnology into crop production and protection, using the same traits that have been used throughout the decades and incorporate cross breeding and genetic modifications. Techniques are being used to increase the size of the yield of the crops, make crops resistant to disease caused by bacteria, fungi and viruses and give them the ability to better withstand harsher elements. Biotechnology has also given us a better understanding of how we can make better use of the plants built in defence systems and it has opened up new avenues allowing us to work with nature by providing us with new bio pesticides. It will serve as invaluable source of references to the scientists, students and teachers working in public and private sectors in the discipline of agriculture biotechnology.

## ABOUT THE AUTHOR

**Dr. Uday Kumar**, Presently Asst. Professor at Janki Devi Mahavidyalaya, Darihat under B. R. Ambedkar University, obtained all his degrees M.Sc and Ph.D. in biotechnology. He is teaching this subject for last thirteen years, since 2005. He has published three research papers and attended Nine National level seminars and presented research papers. His dynamism is also seen in various other fields apart from his successful academic career.

# Biotechnology and its Application in Agricultural Science

Dr. Uday Kumar

AMIGA PRESS. INC
Delhi -110051

**AMIGA PRESS INC.**

H.O :- 28-A, Indra Park,

Chander Nagar, Delhi :- 110051 (India)

Tel :- 011-22504028, +91-98736-57000

Email :- amigapress@gmail.com
amigaeditorial@gmail.com
amigapub@gmail.com

Website :- www.amigapress.co.in

This edition is published by Amiga Press Inc.

ISBN : 978-93-87295-40-7



First Published 2019



---

Printed and bound by Replika Press Pvt. Ltd.

# Preface

Agricultural biotechnology is a collection of scientific techniques, including genetic engineering, that are used to create, improve, or modify plants, animals and microorganisms. Agriculture refers to the production of food and goods through farming and forestry. Agriculture was the key development that led to the rise of civilization, with the husbandry of domesticated animals and plants creating food surpluses that enabled the development of more densely populated and stratified societies. The study of agriculture is known as agricultural science.

The green revolution which gave us plenty of grains to feed millions of people and the revolution in medicine which increased the life span of man are common knowledge even to lay people. All this was possible due to major discoveries and technological innovations in agriculture and medicine. Today, we are witnessing another revolution in biosciences be-cause of some major advances in cell biology and genetics. Many people are inclined to believe that while the battle for green revolution of the type we saw three decades ago was fought in the field, the battle for the new revolution in biosciences known as "Biotechnology" is being fought in modem laboratories.

Modern biotechnology R&D has been conducted in an institutional and economic environment that differs significantly from the development of the earlier Green Revolution technologies. While the latter were essentially the prerogative of public research institutions and philanthropic foundations, developments in biotechnology have been driven essentially as a competitive, commercial endeavor in which powerful private sector actors compete. The major socioeconomic risk of agricultural biotechnology stems from the fact that the research, development, commercialization, and distribution of new biotechnological products have been carried out mainly in developed countries by a few, large, multinational companies.

These companies have focused on temperate crops for large farmers in developed countries. Undertaking R&D on Asia's basic food crops for small farmers in rainfed and marginal areas is of little interest because they see limited returns from such investments. If this trend continues, modern biotechnology will aggravate the income disparity between developed and developing countries, and between large and small farmers. Increased public investments in agricultural biotechnology are necessary to ensure that small, poor farmers have access to biotechnology. Governments must also address the potential gender disparity in the access to technology, and the negative impact specific to women, since no technology is gender neutral. Lessons from earlier agricultural technological changes should be used as a caveat.

The significance of the application of modern biotechnology in the area of food production and its resultant impact in terms of human health and development can not be undermined. As the world is faced with ever increasing population and more and more food shortage and regional imbalances, new technologies and techniques are being developed to enhance production and increase the shelf life of perishable items. It is in this direction that new research initiatives in the field of green biotechnology are being made to enhance productivity and nutrition value of food items.

Green biotechnology which is more commonly known as Plant Biotechnology is a rapidly expanding field within Modern biotechnology. It basically involves the introduction of foreign genes into economically important plant species, resulting in crop improvement and the production of novel products in plants. Use of environment friendly and cost effective alternatives to industrial chemicals such as bio fuels, bio fertilizers and bio pesticides are not only resulting in enhanced crop output, improvement in health and safety standards, these new products are also leading to less environment pollution and use of green technology. The ever increasing demand of agricultural produce has given new impetus to research in the field and has resulted in great benefits for farmers and users alike.

It will serve as invaluable source of references to the scientists, students and teachers working in public and private sectors in the discipline of agriculture biotechnology.

*—Editor*

# Contents

# 1

# Introduction

Technological change in agriculture has been the foundation of economic growth and development. By raising yields and continually increasing the number of people who can be fed by the output of one worker in agriculture, change in agricultural technology has permitted an increasing proportion of labour and other productive resources to be devoted to non-agricultural production. Plant and animal breeding and chemical and mechanical technology plus improved husbandry have caused continual structural change in farming.

These technological stimuli have arisen as a series of overlapping waves to create a process of technological change, the momentum of which looks likely to be maintained by intensified adoption of information and biotechnologies. As a result of the cumulative processes to date, farms have become larger and more specialized, and farmers are more highly trained and have substituted machines for animal and human power; farming in the 'West' has become one of the most capital-intensive of industries.

At the same time agriculture has become more dependent upon industry, and even more industrialized-the intensive production of eggs, broiler chickens, horticultural products and pigs has many characteristics which belong to industrial production rather than to traditional farming. Goodman *et al.* talk of 'the industrial appropriation of the rural processes'. Processes of marketing farm products have passed off the farm to be performed by sophisticated distribution and retailing sectors which increasingly dictate details of production to the farm sector, while at the other end of the chain an increasing proportion of inputs is supplied by the chemical, pharmaceutical and engineering industries.

The process of 'industrializing' agriculture has been transferred to the less-developed countries (LDCs) and is epitomized by the so-called 'Green Revolution'. This has entailed expansion of irrigation systems using modern pumps, engineered dams and canals, plus the increased use of inorganic fertilizer, insecticides, herbicides, fungicides and power machinery; although at its heart has been the 'old' biotechnology of breeding new cereal varieties which give high yields when supplied with chemical inputs. Because much of the impetus for this 'revolution' has been western technology and farming know-how, particularly that of the USA, there are those who are critical of what they see as the LDCs' increasing technological dependence upon the West, its industries and research institutions.

What is undeniably the case is that technological advance has been faster, and public and private expenditure on research in western agriculture greater than in LDCs taken as a whole. Thus exportable surpluses from the West have increased with generous public support, while many LDCs have had to resort increasingly to food imports to meet their needs.

While at this stage it is difficult to foresee exactly when, and on what commercial scale, the new agricultural biotechnology 'revolution' which is now brewing will have its effect, most commentators agree that it will intensify the industrialization of agriculture, and that it will increase the technological dependence of most LDCs upon western firms. It may cause considerable disruption to the economies of some countries and will cause farming operations to diverge increasingly from the pattern associated with the country yeoman. Rural life in many more remote farming areas will be further threatened by the continued deterioration in the economics of ruminant livestock farming on upland and low-productivity pastures. It is these redistributive and structural effects that are the central focus here rather than any attempt to estimate the scale and speed of uptake of specific biotechnology inputs and processes.

Although prospective biotechnological change in agriculture is not different in many fundamental respects from other processes of technological change, in that it will operate through the provision of new or modified inputs (seeds, hormones, etc.) and the creation of new markets in industry, it does raise new issues. Are there dangers from releasing life forms which are engineered using new methods which are quite different from those associated with varieties developed by traditional plant-and animal-breeding methods? Even if the dangers can be objectively assessed and turn out to be minimal, will what may be perceived as man-made and therefore 'unnatural' products be acceptable to society as a whole?

## SIGNIFICANT OF AGRICULTURAL BIOTECHNOLOGY

Every year, millions of people worldwide die from starvation and nutritional deficiencies. Although these challenges remain significant, we are making some progress. Agricultural technologies have increased caloric intake on a per-capita basis in nearly every nation of the world. technological development— particularly in the area of agricultural biotechnology — is crucial if we want to further reduce malnutrition and starvation and meet the needs of growing populations. Unfortunately, misperceptions about biotechnology have led to activist movements to slow down, if not eliminate, this critical technology via government regulation. In the United States, the advancement of biotech crops has slowed down because of attempts to regulate them differently than conventionally grown foods. In Europe, antibiotech movements are much stronger, and international bureaucrats are pushing various international agreements and trade restrictions. While Western nations may be able to absorb the costs of such policies, people in the developing world will pay a higher price as many continue to suffer from starvation.

## What is agricultural biotechnology?

Broadly speaking, biotechnology is any technique that uses living organisms or substances from these organisms to make or modify a product for a practical purpose. Biotechnology can be applied to all classes of organism-from viruses and bacteria to plants and animals-and it is becoming a major feature of modern medicine, agriculture and industry. Modern agricultural biotechnology includes a range of tools that scientists employ to understand and manipulate the genetic make-up of organisms for use in the production or processing of agricultural products.

Some applications of biotechnology, such as fermentation and brewing, have been used for millennia. Other applications are newer but also well established. For example, micro-organisms have been used for decades as living factories for the production of life-saving antibiotics including penicillin, from the fungus Penicillium, and streptomycin from the bacterium Streptomyces. Modern detergents rely on enzymes produced via biotechnology, hard cheese production largely relies on rennet produced by biotech yeast and human insulin for diabetics is now produced using biotechnology.

Biotechnology is being used to address problems in all areas of agricultural production and processing. This includes plant breeding to raise and stabilize yields; to improve resistance to pests, diseases and abiotic stresses such as drought and cold; and to enhance the nutritional content of foods. Biotechnology is being used to develop low-cost disease-free planting materials for crops such as cassava, banana and potato and is creating new tools for the diagnosis and treatment of plant and animal diseases and for the measurement and conservation of genetic resources. Biotechnology is being used to speed up breeding programmes for plants, livestock and fish and to extend the range of traits that can be addressed. Animal feeds and feeding practices are being changed by biotechnology to improve animal nutrition and to reduce environmental waste. Biotechnology is used in disease diagnostics and for the production of vaccines against animal diseases. Clearly, biotechnology is more than genetic engineering. Indeed, some of the least controversial aspects of agricultural biotechnology are potentially the most powerful and the most beneficial for the poor. Genomics, for example, is revolutionizing our understanding of the ways genes, cells, organisms and ecosystems function and is opening new horizons for marker-assisted breeding and genetic resource management. At the same time, genetic engineering is a very powerful tool whose role should be carefully evaluated. It is important to understand how biotechnology-particularly genetic engineering-complements and extends other approaches if sensible decisions are to be made about its use.

## Understanding, characterizing and managing genetic resources

Farmers and pastoralists have manipulated the genetic make-up of plants and animals since agriculture began more than 10 000 years ago. Farmers managed the process of domestication over millennia, through many cycles of selection of the best adapted individuals. This exploitation of the natural variation in biological organisms has given us the crops, plantation trees, farm animals and farmed fish of today, which often differ radically from their early ancestors.

The aim of modern breeders is the same as that of early farmers-to produce superior crops or animals. Conventional breeding, relying on the application of classic genetic principles based on the phenotype or physical characteristics of the organism concerned, has been very successful in introducing desirable traits into crop cultivars or livestock breeds from domesticated or wild relatives or mutants. In a conventional cross, whereby each parent donates half the genetic make-up of the progeny, undesirable traits may be passed on along with the desirable ones, and these undesirable traits may then have to be eliminated through successive generations of breeding. With each generation, the progeny must be tested for its growth characteristics as well as its nutritional and processing traits. Many generations may be required before the desired combination of traits is found, and time lags may be very long, especially for perennial crops such as trees and some species of livestock. Such phenotype-based selection is thus a slow, demanding process and is expensive in terms of both time and money. Biotechnology can make the application of conventional breeding methods more efficient.

## Genomics

The most significant breakthroughs in agricultural biotechnology are coming from research into the structure of genomes and the genetic mechanisms behind economically important traits. The rapidly progressing discipline of genomics is providing information on the identity, location, impact and function of genes affecting such traits-knowledge that will increasingly drive the application of biotechnology in all agricultural sectors. Genomics sets the foundation for post-genomics activities, including new disciplines such as proteomics and metabolomics to generate knowledge on gene and protein structure, as well as their functions and interactions. These disciplines seek to understand systematically the molecular biology of organisms for their practical use. A vast range of new and rapidly advancing technologies and equipment has also been developed to generate and process information about the structure and function of biological systems. The use and organization of this information is called bioinformatics. Advances in bioinformatics may allow the prediction of gene function from gene sequence data: from a listing of an organism's genes, it will become possible to build a theoretical framework of its biology. The comparison across organisms of physical and genetic maps and DNA sequences will significantly reduce the time needed to identify and select potentially useful genes.

Through the production of genetic maps that provide the precise location and sequences of genes, it is apparent that even distantly related genomes share common features. Comparative genomics assists in the understanding of many genomes based on the intensive study of just a few.

For instance, the rice genome sequence is useful for studying the genomes of other cereals with which it shares features according to its degree of relatedness, and the mouse and malaria genomes provide models for livestock and some of the diseases that affect them. There are now model species for most types of crops, livestock and diseases and knowledge of their genomes is accumulating rapidly.

## CONTROL OF BIOTECHNOLOGY DEVELOPMENT

The role of the public sector to undertake research and development of agricultural technology as a public service to firms which might profit from translating that R & D into a commercial product or process, to farmers who might profit from adoption of the technology, and perhaps most importantly to consumers at home and abroad who benefited from the lower prices resulting from greater abundance. This was particularly true of the phase of change dominated by improvements in plant and animal breeding, the 'old' biotechnology. While the public sector had a role in basic research for chemical and mechanical technologies, the benefits of research expenditure in these areas were easier to capture by private companies investing in R & D, so that increasingly the public sector has taken a smaller role in these areas although maintaining a strong regulatory role in regard to agricultural chemicals in particular. Where it is impossible to prevent others from escaping payment for the research costs, either because the product is easily copied (seeds which can be regenerated by farmers or other firms) or because proposals for new methods can be readily implemented, private firms are understandably unwilling to invest.

Machines, insecticides, fungicides and other manufactured inputs do lend themselves more readily to private exploitation. Nevertheless the returns to R & D in these products do depend upon the degree of difficulty potential competitors would have in copying the product. In some cases there are inherent technical difficulties in copying the process, or it would be prohibitively expensive, but in other cases it is the ability to obtain patent protection which creates legal barriers to potential competitors' ability to become 'free-riders' and which protects incentives to private investment in R & D. Traditionally, however, it has been impossible for plant and animal breeders to obtain patent rights for their products, which is one reason why public-sector R & D has remained so important in this area.

For, as under the European Patent Convention (EPC) of 1973, it has been judged impossible for plant varieties to satisfy one of the key criteria to qualify for a patent, namely proof of 'an inventive step'. The application of standard breeding practices to generate new varieties by crossing existing plants or animal strains has not been deemed to be invention.

Thus under the EPC one set of exclusions from patentability is 'Plant or animal varieties or essentially biological processes for microbiological processes or the products thereof.' In the absence of patent rights plant-breeding firms in particular have worked hard to obtain other means of protection and royalties for their products.

The history of the development of Plant Breeders' Rights (PER) is presented by Mooney, and is of particular interest because of the concerns he expresses about the consequences of allowing the basic genetic stock, which underpins agriculture and hence our whole society, from becoming private rather than public property. While Mooney's concerns are expressed in relation to 'old' biotechnology varieties of plants, they are of particular importance with respect to the products of the 'new' biotechnology. For bio-engineered products are capable of meeting the inventive-step criterion of patentability, and both micro-organisms and transgenic animals have now been patented in the USA.

Before examining developments in the seed industry it is worth touching upon the implications of changes in research policy which stress increasing reliance upon private R & D to develop and exploit the new biotechnology and other technologies. In the UK government has decided that it should withdraw from funding what it terms 'near-market research', that is, research beyond the basic phase and which is preparation for commercial exploitation. This policy is based in part upon the arguments that industry should be investing more heavily in R & D, and that research strategies at the near-market stage should be driven by assessments of likely commercial success which can best be made by the firms involved, and will therefore lead to greater efficiency in allocating research funds.

This has led to the closure of a number of government-financed agricultural research institutes, the scaling down of others and the sale of the National Seeds Organization by auction to Unilever; Unilever won against competition with BP and other major public companies. This deliberate attempt to switch an increasing proportion of agricultural R & D expenditure from the public to the private sector carries with it a number of risks.

In the first place there is a controversy about whether there is under-investment in R & D so that the returns to extra expenditure are high, or whether the converse is the case. If there is under-investment, then withdrawal of public support for applied research will exacerbate this, as the private sector will only undertake R & D expenditure on those products and processes from which exclusive benefits can be captured by the investor.

Thus in relation to biotechnology in the USA, where again public support for applied biotechnology research is relatively small, Stallman and Schmid argue that there will be emphasis on technologies which are applied in factory conditions where secrecy and control can be maintained. For technologies which cannot be confined to factories, such as seeds, Stallman and Schmid state that 'Firms are also considering mechanisms which "scramble" the genome of a plant in the second generation', in order to prevent farmers from reproducing seed with the enhanced, engineered characteristics.

Clearly such actions are designed to frustrate the maximum spread of benefits from the technology and to maximize private profit for companies investing in research.

Another facet of commercially orientated biotechnology research is that it will aim at the most important crops and livestock products, those produced by the largest farming units, and those which are most heavily subsidized. In the latter case this will worsen the budgetary problems of adjusting agricultural policy in OECD countries. The probable neglect of minor crops, difficult habitats and small farmers means that the public sector will have a defined role in agricultural technology research, but one in which it will be relegated to the second division and where it is unlikely to prove successful in terms of the commercial yardstick of rates of return which is increasingly emphasized by public-research policy.

## GENETICALLY ENGINEERED PLANTS

At a symposium at the University of Washington Law School in October 1993, when discussing the 'Future of Intellectual Property Protection for Biotechnology in the US, EC, and Japan', Joseph Straus of the Max Planck Institute (Germany) warned the participants against 'ethics and other irrational considerations'. Indeed, except for a very narrow band of 'moral dilemma' situations which are the stock-in-trade of professional biomedical ethicists, the ethical aspects of genetic engineering have been routinely ignored by government policy-makers and corporate technology promoters. My thesis is that it has been the corporate promoters and their governmental handmaidens who have been 'irrational' in their systematic refusal to acknowledge the environmental and ethical considerations of genetic manipulation.

Technologies, by definition, are neither acts of God nor nature; they are the embodiment of specific human purposes and intentionality. Far from being inevitable, they are researched and developed by those entities with sufficient power to mobilize social institutions to bring the technology into being, according to their own goals and normative considerations. 'Biotechnology is not neutral. It shares the propensity of modern materialistic science to desacralise, dominate and manipulate life. It reduces all living things to a mechanism which it can manipulate according to engineering standards.' Genetic engineering is a technological process for undertaking activities which do not, and cannot, occur in nature. In this sense it is perfectly legitimate, therefore, to label genetic engineering as 'unnatural'. The European Community has incorporated this concept in its definition of genetically modified organisms, and it has been adopted by the UN Environment Programm (in the working of its Fourth Expert Panel set up under the Biodiversity Convention as part of the follow through to the United Nations Conference on the Environment and Development held in Rio de Janeiro in June 1992): 'organisms in which the genetic material has been altered in a way that does not occur naturally by mating and/or natural recombination'.

Thus, hybrids and other modified organisms obtained by traditional breeding techniques are excluded from the definition of genetically modified organisms. It is important to note that the process as well as the products are novel. The promoters of the new technology believe that a concern for the process itself is specious, downplaying its novelty whenever confronted by discussions of regulatory oversight. However, one of the bases of the decision of the US Supreme Court in the *Chakrabarti* case (which, 5 to 4, upheld the patentability of genetically engineered microbes, and which has been unquestionably accepted by foreign governments as disposative of the issue of patentability of genetically engineered life forms) was that the element of 'novelty' required by the patent law was to be found in the process by which the microbe had been created. One reason why we cannot ignore the powerful novel aspects of the processes of genetic manipulation is because we are not omniscient as to what will, in fact, happen when we alter genomes.

Genetic manipulation is not like a child's game of Legos or Tinkertoy, in which parts can be rearranged or linked up with only simple mechanical and relatively predictable consequences. In calculating any risk from a transgenic organism, one should consider four elements: the host organism, the foreign genes, the interaction between the foreign genes and the rest of the genome, and the environment in which the organisms will be used.... [in regard to the last two elements] the literature contains many examples of genetic manipulations where inserted genes did not respond in their new environments the way they did in their old ones or where alterations with one part of the genome caused surprising activity in other parts of the genome. There is an element of arrogance to scientific assertions which assure us that the process itself poses no risks, since scientists know so little about actual ecosystems.

For example, I have been told by agronomists that over 80 per cent of the organisms which can be identified in a soil sample from my garden are completely unknown to the scientific literature. Indeed, the Ecological Society of America itself has specifically warned about the problems presented in our lack of ecological knowledge and the resulting consequences from releasing (intentionally or accidentally) genetically altered organisms into ecological systems.

## Foods

The Food and Drug Administration, responding to corporate pressures to remove the prospect of regulation of genetically altered foods, issued a set of rules which have largely left responsibility for protecting the public health and safety in the hands of the industry, permitting products to be marketed without scrutiny unless the industry indicated to the agency that it believed governmental oversight was justifiable. However, the proposal as published by the FDA in the Federal Register perversely noted several important problem areas: creating allergens, additions of genes from sources which might violate religious and cultural norms (of vegetarians, Jews or Muslims, etc.), and implications for animal welfare and well-being (for example, the incorporation of human growth hormone-producing gene into a pig's genome produced a highly arthritic animal).

Indeed, the commissioner of the FDA, David A. Kessler, and his colleagues noted the possibility that genetically engineered food might 'contain high levels of unexpected, acutely toxic substances'. In order to begin to address ethical issues presented by the genetic modification of foodstuffs, we need to have some ideas of the purposes for which these modifications are performed. (For example, if we wish to approach an analysis on utilitarian cost-benefit grounds.) Despite a great number of general statements, human nutrition and hunger do not appear to be the actual driving forces behind the development of genetically engineered foods. Other than claiming increased shelf life (which can be indirectly related to nutrition in the sense of reducing spoilage or the consumption of foods that have begun to spoil), there is hardly any indication at all that genetic engineering is being directed to create nutritious substances out of non-nutritious ones.

There are few real instances of genetic engineering reducing the cost of foodstuffs, or increasing the quantities of foods actually available to populations that are hungry because of the non-existence of consumables (as opposed to hungry because they lack money to buy sufficient foods, no matter how produced). Indeed, early genetic manipulations seemed perversely designed to minimize the achievement of such goals: the creation of 'herbicide tolerant plants' which do not reduce the applications of dangerous chemicals to agricultural fields (which might occur if genetic manipulations of food crops were designed to make them resistant to insects, fungi, or disease) but instead permit higher levels of chemical application, or the introduction of recombinant bovine growth hormone to produce more milk at a time when developed countries suffer from milk gluts and have instituted programmes to kill cows and physically dump milk (milk as a commodity is price inelastic; quantity increases do not lead to price decreases).

The goals of genetically engineering foods seem to be more closely tied to increasing the economic gain and power of the corporations involved in food production (for example, the development of herbicide-tolerant farm crops has been led by corporations which manufacture herbicides), making production easier for corporations, reducing or altering packaging and transportation costs for the corporations, etc.

## *Environment*

Industrial societies around the world are faced by massive pollution legacies from the previous emphases on chemical and nuclear industrial activities. The inability of the US Department of Energy to find a suitable site for a longterm nuclear waste repository illustrates our very poor track record for dealing with the scientific, technical, economic, and socio-political aspects of prior technological 'revolutions'.

The environmental problems posed by genetically engineered organisms are likely to be substantially more intractable than those posed by these earlier instances of pollution because genetic wastes multiply, migrate, and mutate. A genetically engineered organism once free in the environment is impossible to recall. (Impacts which are irreversible must be considered with higher scrutiny than those which can be undone.) Examples of environmental problems which need to be assessed include the risks of transgenic crops themselves becoming weeds; the risk of gene flow to wild relatives which might become weeds or pests; the growth of antibiotic resistance in species (particularly animals), since resistance genes are used as markers for the genetic engineering sites; the problem of exotic or non-native species taking over ecological niches (such as gypsy moth, starling, kudzu, rabbits in Australia, etc.) leading to extraordinary economic losses as well as ecological ones.

The restriction (rather than increase) of biodiversity by selective advantageous breeding of transgenic organisms as opposed to natural ones, or the predatory results of expanding exotic species (such as the Dutch Elm micro-organism destroying the American Elm and severely restricting the biodiversity of certain areas in the Northeast United States); health issues (both of plant and animal species, as well as humans — worker health and safety as well as community security); and the occurrence of the completely unexpected, the inability to eliminate uncertainty

For example, the late 1993 floods in the Mississippi valley included the flooding of a field of genetically engineered corn and the dispersal of this plant material to unknown sites within the thousands of square miles of downstream flood plain. Prospective ecological assessment is not being performed, despite the fact that its need has long since been recognized. And economic priorities, particularly those accruing directly to the promoters, as well as the nebulous spur of 'competition' with foreign countries, oftentimes overwhelms any interest in doing environmental impact analysis.

## DEVELOPMENTS IN THE SEED INDUSTRY

Many major food crops upon which we depend originated and were first cultivated in Third World countries rather than in the industrialized countries where they are most productively exploited today and in which they have been improved by traditional plant breeding. The potato originated in South America, wheat in Ethiopia and the Near East, important maize varieties in South America, and rice in Asia. Moreover the so-called Vavilov centres, which contain the greatest variety of living species and which have been the source of much important gene material, are mainly in Third World countries. The wealth of the industrial countries owes much to their past ability to exploit agriculturally crops and animals originating in other countries.

In the colonial era the success of British scientists in smuggling rubber plants from Brazil to Sri Lanka, Singapore and Malaysia and the comparable more complex route by which coffee was introduced to Latin America from Ethiopia are instances of colonial powers obtaining significant wealth by exploiting plant material from the Third World. In one way or another the process by which institutions and firms in developed countries have continued to collect new plant and animal varieties from less-developed countries has continued. But what is still a major source of friction is the procedures by which legislative protection is developed to give companies in the richer countries commercial rights to exclusive exploitation of varieties developed from plants freely collected from other countries, and to extract high rates of profit from the sale of those varieties. Most extreme is the situation whereby under the USA Plant Patent Act of 1930, which covered asexually propagated plants such as certain fruits, flowers, ornamental shrubs and trees, it has been possible for plants discovered and smuggled out of other countries to be given patent protection in the USA. The key to this is that the Act accepted the unfeasibility of requiring proof of an 'inventive-step' for the granting of a patent, and employed the criterion that as far as can be determined the plant variety is new.

Since some novel variety found in the rain forests of South America may satisfy this criterion, and the source of the plant can be concealed from the authorities, the theft and smuggling of plants can be said to have been encouraged. Seed companies in developed countries have successfully pressed their case for protection, and have progressively extended Plant Breeders' Rights (PBR) through both national legislation and the Union for the Protection of New Varieties of Plant (UPOV), which is an international agreement among signatory countries to honour a system of varietal rights.

In the case of food crops, in the absence of patentability, PBR operates by registering distinct varieties. Pressure has been exerted upon less-developed countries to become signatories of UPOV and thus acknowledge the rights of breeders, most of whom are from developed countries. However, the majority have stalled or rejected membership, and have been fundamentally unhappy about accepting the principle of creating monopoly rights for others over plant material which in some cases originated in their own country.

The whole issue of intellectual property rights has been raised to a prominent political level by the USA's insistence that it be included as one of the thirteen areas for negotiation in the Uruguay Round of multilateral bargaining of the General Agreement on Trade and Tariffs. While the issue is a much broader one than that of plant and animal breeders' rights, it reflects a basic political conflict between the richer countries, determined to obtain secure returns on R & D expenditures by companies, and LDCs struggling to catch up and anxious to avoid having to pay royalties on products protected by PBR and patents. This conflict has broader moral and ethical dimensions when it relates to the issue of charging royalties for the seeds of food crops to countries where malnutrition is widespread.

Against the previous argument has to be balanced the need of those incurring R&D expenditure to capture enough of the returns to provide an incentive for R & D. At least that is certainly the case where R & D is to be undertaken by the private sector, which is politically the increasingly preferred solution in the UK, the USA and elsewhere. This, however, leads to another area of concern, which is that a small number of large multinational chemical and pharmaceutical firms have increasingly come to control the seed industry. Companies such as Shell, Sandoz, Dekallb-Pfizer, Ciba-Geigy and BP have become owners of most of what were small independent seed companies.

This linking between chemicals and seeds within firms has a strong commercial logic. It facilitates the marketing of packages consisting of seed varieties plus the appropri-ate fertilizers, insecticides, etc. It also allows the companies concerned to integrate their plant breeding, biotechnology and chemical research. One particular synergy for these companies will be to bioengineer varieties of major food crops to resist their own herbicides and weedkillers. This would permit chemical weed control to be extended to large-scale crops where this is not now possible, and is just one way in which the direction of biotechnolo-gical research may be influenced (biased) by the combination of forces giving a small number of companies extensive control over the basic means of agricultural production. There are many other issues which could be touched on regarding corporate influence over the development of agricultural biotechnology. It is, however, understandable that the spokesmen for LDC countries, which are profoundly concerned about their dependence upon western technology and exercised by what is seen as colonial and post-colonial exploitation by the West, should be alarmed by the prospect of new rights being created over seeds and other biotechnological products. Certainly there are widespread fears that they will only gain access to the fruits of biotechnology on unfavourable terms and that their relative dependence on and subservience to western industry will be increased so that benefits to them will be small. There are many western critics who would agree with this, and that the system by which agricultural biotechnology is delivered will favour the 'haves' rather than the 'have-nots'.

## AGRICULTURAL BIOTECHNOLOGY REGULATION

For traditionally bred plants, regulators rely on plant breeders to conduct appropriate safety testing and to be the first line of defence against genetic alterations that might prove dangerous. The same should be true for agricultural products developed using biotechnology. There is no reason for regulators to treat biotech plants any differently than traditionally bred plants, particularly given the fact that biotechnology provides greater control over gene manipulation. Despite fears about gene manipulation, traditional crossbreeding has altered plant genes and improved the human diet for the past 30 years. In fact, people worldwide safely consume these plants every day. Examples include wheat, corn, potatoes, tomatoes, and countless other staples of the American diet. In 1984, the biotechnology industry began experimenting with "gene-spliced plants" — the more advanced approach to gene manipulation that we now call "biotechnology."

### Department of Agriculture

USDA regulates the release of all genetically engineered agricultural plants under statutes giving USDA's Animal and Plant Health Inspection Service (APHIS) the authority to regulate plants that may be or may become weeds or other nuisances — what the statutes call "plant pests." Although the rules apply in a general sense to novel or exotic varieties of both gene-spliced and conventional plants, APHIS typically only requires field testing of conventional plants that are new to a particular U.S. ecosystem (transplanted from another continent, for example).

However, all genetically engineered agricultural plants face a higher regulatory bar — making it more difficult to expand the use of biotechnology. Genetically engineered crops must be field tested under APHIS regulations prior to commercialization. For example, a new variety of corn produced with conventional hybridization requires no government-mandated field testing, but all new varieties of genetically engineered corn do, even though there is no logical reason for the regulatory disparity. For most genetically engineered plants, APHIS requires the company producing the plants to submit notice detailing the gene or genes that have been inserted, where the plant will be tested, and other relevant characteristics of the plant prior to receiving permission to conduct the field trials. Once the company completes field testing, APHIS reviews the results and makes a determination on whether or not the product should be "deregulated" and can be released into the market.

## Regulatory Scheme

At the time, the White House Office of Science and Technology Policy began crafting a framework wherein existing federal agencies would regulate genetically engineered organisms on the basis of their characteristics, not the method of production — thus wisely deferring to the scientific community's judgement that regulation ought to address the products, not the process of biotechnology. Engineered organisms would not require extra scrutiny simply because genetic engineering produced them (the process). Instead, they would be subject to heightened scrutiny only if the individual organisms expressed characteristics that posed some conceptually heightened risk (the product). The federal government divided regulatory jurisdiction among agencies already involved in agricultural, food, and environmental regulation. These include the United States Department of Agriculture (USDA), the Environmental Protection Agency (EPA), and the Food and Drug Administration (FDA). While each of these agencies considers the characteristics of individual products in their regulation, only FDA followed the general scientific thinking that genetically engineered and non-genetically engineered products should be regulated similarly. Both USDA and EPA automatically subject all genetically engineered plants as a class to premarket approval requirements not ordinarily applied to conventionally bred plants.

## Environmental Protection Agency

EPA regulates plants that are modified to produce substances that act like pesticides: that is, substances used by a plant to protect itself from pests, such as insects, viruses, and fungi. EPA's proposed rule for these "plant-incorporated protectants" is not yet finalized. In the interim, however, EPA has regulated such plants by applying its proposed guidelines, which are functionally similar to rules for the registration of synthetic chemical pesticides. Again, biotech products face higher regulatory hurdles, even though plant-incorporated protectants developed through conventional breeding are exempt from these requirements.

## Health and Human Services: Food and Drug Administration

FDA is responsible for ensuring that food items, including foods derived from genetically modified plants, are safe to eat. Following the general regulatory framework that emphasizes product regulation rather than process regulation, FDA rightly does not treat foods derived from genetically engineered plants to be inherently unsafe. Food producers are not required to seek premarket approval from FDA unless there is a substantive reason to believe that the novel trait (or traits) in the food pose a safety question. The initial determination of safety is left to the producer, but FDA has encouraged producers to consult with agency scientists prior to marketing a food produced with biotechnology to ensure that the appropriate determination is made. Recently, FDA published a proposed rule that would require producers to notify the agency at least 120 days before marketing, which may be a signal that FDA could abandon its more reasonable approach in the future. In addition, FDA does not require labelling of foods derived from biotechnology unless the genetic insertions so alter the food that the common name no longer describes it adequately. Examples of this phenomenon would include such alterations of the food product that would raise a safety issue or change the product's nutritional composition or its storage or preparation characteristics.

## International Trade

While U.S. consumers do not appear to be strongly opposed to biotech foods (they, in fact, seem rather indifferent), a strong anti-biotechnology movement has arisen in several European countries over the last few years. The European Union (EU) has established strong restrictions on the commercial planting of genetically engineered plants, and European food processors and retailers are reluctant to import harvested agricultural products derived from biotechnology — thus jeopardizing the marketability of U.S. commodity grain exports. And the EU is now negotiating for strong restrictions on agricultural biotech products in important international agreements governing food safety and environmental protection. Very strong restrictions were included in the Biosafety Protocol, finalized in January 2000.

Perhaps, more important, though, are the Codex Alimentarius Commission standards for food safety. Without convincing scientific evidence that genetically engineered crop plants pose a heightened environmental or human health risk, restrictions on agricultural biotech imports could be challenged under the General Agreement on Tariffs and Trade and the World Trade Organization. Thus the EU has resorted to justifying its decisions on the basis of a risk-management philosophy known as the "precautionary principle," and the EU advocates inclusion of the precautionary principle in international environmental and food safety agreements (such as the Codex Alimentarius), as well as within GATT itself. Its inclusion would give nations greater latitude in restricting imports of U.S. agricultural products.

No single definition exists for the precautionary principle, but its general meaning is that when an activity raises threats of harm to human health or the environment, regulatory measures should be taken to prevent or restrict the activity even if the risks have not been demonstrated scientifically.

Although the EU asserts that the precautionary principle is an unbiased risk-management philosophy, critics have noted that its lack of definition and evidentiary standards makes it all too easy to abuse for the purpose of masking trade protectionism, and that its very approach to risk management is inherently flawed and may, in fact, increase net risk.. Nevertheless, the precautionary principle has already been included in several international environmental treaties, including the Convention on Biological Diversity and the Biosafety Protocol.

## Labelling Issues

Some activists, however, argue that the government should mandate the labelling of all genetically engineered foods. They assert that consumers have a "right to know" how their foods have been altered, and that a mandatory label would best allow consumers to choose between genetically engineered and conventional foods. Biotechnology advocates have argued against mandatory labelling because such requirements raise food costs— something that mostly harms lower-income Americans and people on fixed budgets. Perhaps more important, while biotech products are not substantially different from other products, special labels would likely make consumers think these products were more dangerous.

Hence, rather than serving educational or "right to know" purposes, such labels promise to simply confuse consumers. A government-mandated label on all genetically engineered foods also would raise important First Amendment free speech issues.

In 1996, the U.S. Court of Appeals, in *International Dairy Foods Association et al. v. Amestoy*, ruled unconstitutional a Vermont statute requiring the labelling of dairy products derived from cows treated with a genetically engineered growth hormone, noting that food labelling cannot be mandated simply because some people would like to have the information. "Absent ... some indication that this information bears on a reasonable concern for human health or safety or some other sufficiently substantial governmental concern, the manufacturers cannot be compelled to disclose it."

In other words, to be constitutional, labelling mandates must be based in science and confined to requiring disclosure of information that is relevant to health or nutrition. Furthermore, consumers need not rely on mandatory labelling of biotech foods to truly have a choice. Real-world examples show that market forces are fully capable of supplying information about process attributes (including kosher and organic production standards) that consumers truly demand. The same can be said about non-biotech foods, and the FDA recently published proposed guidelines to assist producers in voluntarily labelling both genetically engineered and non-genetically engineered foods. Additionally, the USDA's newly published rule for organic certification necessarily excludes biotech products from organic food production. Consequently, consumers wishing to purchase non-biotech foods need only look for certified organic products. Even as underdeveloped nations clamor for biotechnology applications, and as countries like China continue to experiment with and use agricultural biotechnology, opponents of agricultural biotechnology in the West, particularly Europe, attack it as an unnatural process that will destroy the world, not better it. They argue that biotechnology should be heavily regulated, if not banned. Already, however, genetically engineered plants are subject to strict regulatory oversight that is equal to or greater than that advocated by the vast majority of scientific specialists.

Additional regulation will slow down research and development of genetically engineered crops, keep beneficial products off the market, and raise the cost of products that do make it to consumers. Furthermore, the inclusion of similar restrictions — or inclusion of the precautionary principle— in international agreements will greatly impact the international trade of agricultural goods and delay their introduction into the marketplace. Each of these problems could prevent this technology's benefits from being introduced to industrialized nations and, more important, the developing world.

## BIOTECHNOLOGICAL PRINCIPAL OF AGRICULTURE

Agricultural technology can be defined as biotechnology to include 'any technique that uses living organisms or processes to make or modify products, to improve plants or animals or to develop micro-organisms for specific uses- it focuses upon recombinant DNA and cell fusion technologies'. Longworth concurs with this definition, with the significant addition of tissue-culture techniques, an aspect of biotechnology which has great economic, and therefore social and political, potential impact.

Despite debates about whether these or any other definitions are adequate, they do touch on the main aspects of biotechnology which deserve to be considered. The aspect of this new biotechnology which most captures the imagination and stirs the greatest controversy is gene splicing or recombinant DNA techniques, which inspire researchers to consider the possibilities of producing reproducible animals and plants markedly different from current existing species, and referred to as transgenic species. Already in higher animals and plants recombinant DNA has produced transgenic forms which are being commercially exploited. Their agricultural significance is, however, so far limited, and commercial application is concentrated in highly profitable pharmaceutical and horticultural markets. Genes have been introduced into several animal species which alter their protein synthesis to enable transgenic sheep to produce insulin in their milk, and rabbits to produce interferon. A completely different application has originated in Denmark for salmon, where it has proved possible to introduce germplasm which enables the salmon's physiology to handle heavy metals, which are normally toxic, so opening up new locations for farm fisheries. Much current research is directed to conferring disease immunity on animals, and holds out the prospect of widespread commercial application.

In plants one achievement has been the transfer of genetic resistance to antibiotics in the petunia, and another has been the introduction of storage-protein genes from French bean plants into tobacco plants. As yet commercial progress with recombinant-DNA technology in plants appears limited, but extensive opportunities beckon, particularly because plant research is less restricted by the ethical and animal-welfare concerns which apply to research on transgenic animals. The most important commercial developments based on gene splicing have so far occurred with genetically much simpler microorganisms, and it is with these that the greatest short-to mediumrun commercial potential lies.

Already genetically engineered micro-organisms are producing a variety of hormones, vaccines, enzymes and other proteins. Important examples are the production of insulin, vaccines for neo-natal diarrhoea in calves and piglets, and the bovine growth hormone BST identical to that produced in cows which can stimulate a 20 per cent increase in milk yield. BST is already a source of problems for legislators in the European Community, and has provoked strong reactions from the media and milk consumers. As far as larger animals, and cattle in particular, are concerned, it is developments in embryo transfer and in many processes for manipulating reproduction which hold out the prospect of continuing increases in yields, feed conversion efficiency and general economic efficiency.

Already, apparently over 1 per cent of dairy calves in the USA are from embryo transplants, despite the high costs still associated with this procedure. Widespread adoption of these sophisticated technologies would further distance livestock farming from its traditional rural simplicity, and from the natural mating of animals to produce offspring. It would place technical demands upon the operators which favour large-scale company farms capable of supporting a range of highly trained specialists. Longworth identifies new techniques of tissue and cell culture as having 'the potential for enormous advances in crop improvement in the next couple of decades'. Those techniques 'can both increase the genetic diversity and greatly increase selection efficiency', and they permit innumerable plants to be reproduced asexually from single cells or small pieces of tissue.

The particular technique known as 'callus culture' has been used for many years to clone highly valued horticultural plants such as orchids, and cloning of cuttings is widely practised for tree crops such as tea and palm oil as well as by millions of gardeners for garden plants. According to Longworth 'cell culture' of single cells has unexpectedly, and so far inexplicably, resulted in plants with different properties being regenerated from the same clump of parent tissue. Sugar cane, maize and potato plants regenerated in this way have been found which are resistant to important pathogens.

This application of cell culture with the capacity to generate vast numbers of seedlings rapidly has the potential to simplify greatly the hitherto labourious procedures of plant breeding and selection. However, in terms of current and immediate commercial importance, it is through micro-organisms that biotechnology has its greatest impact. Microbial fermentation processes have been of great commercial significance for centuries, for example in bread, wine and cheese production, and there is the prospect of considerable development. Two recent examples of new processes indicate the sorts of impact that such developments can have.

In the late 1960s genetically engineered bacteria were developed which were able to digest corn starch to produce high-fructose corn syrup (HFCS) and which left as a residue corn gluten, which is now an important protein feed for livestock. HFCS has made considerable inroads in the United States and in Japan. It has largely replaced sugar as a sweetener in Coca Cola as well as many other food and drink products. This has helped depress sugar prices to Third World growers. In the European Community steps have been taken to prevent HFCS and other new powerful sweeteners from being produced and from undermining the market for domestically grown beet sugar, which is supported by the Common Agricultural Policy.

A second important application of microbial fermentation has been the production of ethanol from sugar cane in Brazil and from maize in the USA. The commercial viability of these processes is critically dependent upon the price of oil as the main non-renewable source of fuel, and, to date, massive subsidies have been required to maintain the Brazilian and USA ethanol programmes. Longworth states, however, that a new biotechnology, Sucrotech, is being patented, which will not only reduce the cost of producing ethanol from sugar cane, but will simultaneously produce fructose at a cost which will be competitive with HFCS and may thereby reclaim part of the sweetener market for sugar cane.

That such possibilities are in prospect indicates how volatile the future might be; biotechnology has tilted the competitive balance from sugar cane to maize, causing economic pressure and even disruption to sugar-cane-dependent economies and may in future switch it back again. In the long run the capacity to ferment fuels microbially from renewable agricultural feedstocks points to an important long-term reorientation of agriculture if non-renewable oil becomes uncompetitively expensive as a fuel for cars and feedstock for certain chemicals. It suggests that eventually there will be an increased emphasis on agricultural production of industrial feedstocks at the same time as continually increasing food output will be required to feed the expanding world population. All this will require considerable increases in agricultural productivity, to which biotechnology will increasingly contribute, but it will at the same time impose great strains on the natural environment and the structure of agriculture.

## USES OF BIOTECHNOLOGY

Biotechnology-the sum of the technologies deriving from molecular and cell biology-will have utility in plant breeding beyond the prebreeding stages. Genetic transformation—non-sexual insertion of alien genes and/or gene regulating systems—is envisaged as eventually being applied, quickly and easily, to finished cultivars, to endow them with a series of useful new traits.

However, experience is beginning to show that genetic background affects expression of genes and gene systems that are transferred *in vitro,* as is also the case when they are transferred via regular sexual mechanisms.

Long periods of trial and error are needed to bring transformed lines into desired levels of expression. Thus, for some years to come, genetic transformation perhaps will be used only as a prebreeding technique, one in which alien genes are put into breeding pools of elite germplasm, following which standard plant breeding methods are used to extract good new varieties containing the alien genes. Cell culture techniques—particularly protoplast fusion techniques—will have utility for making new combinations of nuclear and cytoplasmic genes.

Immediate utility is seen for manufacturing new male sterility systems to make F1 hybrid cultivars, or to develop new kinds of herbicide resistance. With time and experience it may be possible to increase primary productive potential as well, by selecting and fixing optimum mitochondrial and chloroplast genotypes, balanced with proper nuclear genotypes. Such success seems far in the future, however, since at present little is known about optimization of nuclear, mitochondrial, and chloroplastic combinations. DNA technology will help plant breeding in a more traditional way, when information is further developed on maps of restriction fragment length polymorphisms (RFLPs) and their linkages with useful traits in crop plants.

*In vitro* tests, using RFLP markers, will allow near perfect accuracy in selecting for such useful linked traits. RFLP technology also will be used for precisely characterizing cultivar lineages and relationships, for positive identification of miss-identified or mislabeled cultivars, for characterizing gene frequencies of populations, and even to test hypotheses of landrace evolution, and global movement of crop genotypes. Perhaps the greatest contribution of molecular biology to plant breeding will be a deeper and more precise understanding of gene action and inheritance, of cellular physiology and biochemistry, and perhaps of whole plant physiology. With such deepened understanding, breeding and selection via standard sexual methods will be planned and executed with greater vision and economy. Such knowledge and execution will be, in part, a consequence of efforts to make genetic transformnation a practical reality. The fall-out of knowledge from genetic engineering efforts, wisely utilized by observant biologists, will be of great benefit to plant breeding, and to plant culture in general.

## DEVELOPMENT OF NEW CROPS

In addition to broadening the genetic base of established crops, genetic enhancement-and plant breeding in the sense of final cultivar development-can be used in two other ways: to develop new crops from heretofore uncultivated species, and to change old crops into new crops. Breeding of the first type (creating new crops where none existed before) usually is interpreted also to include development of advanced cultivars from landraces.

Usually the advanced cultivars are made for industrial countries and the landraces come from developing countries. Soybeans and amaranth thus are spoken of as "new crops" in the United States, even though they have been cultivated for centuries as landraces in China, and in Central and South America, respectively. Challenges are great and are about the same, for changing either wild species or landraces into advanced cultivars, able to withstand monocropping and intensive culture at high production levels. Selection for productivity inevitably requires narrowing the germplasm base, in the course of selecting plant architecture and physiology able to withstand crowding and the stress of high yields, and amenable to machine planting, weeding and harvesting. Planting the narrowly selected genotypes in large areas gives excellent access to specially adapted disease and insect pests, as already noted, and so the need for tolerance or resistance always is next to be added to the list of breeding priorities for new crops. Thus, genetic enhancement first narrows and then later must broaden the genetic base, in the course of new crop development. And as soon as users of the new product get used to larger quantities of it, they will want to standardize and then to change the quality of the product-as, in oil percentage, grain storability and hardness, or flavour and texture. The rule is simple: the more successful the crop, economically, the greater the demands on (and complaints about) the plant breeders.

Breeding of the second type-changing old crops into new ones-has been considered as a broad-scale possibility only since concepts of genetic engineering were developed. Ability to produce human insulin in bacteria gave rise to thoughts of producing insulin, or other useful primary gene products, in crop plants. Plants were envisaged as now able to turn sunshine, air and water (and soil nutrients) into high value specialty chemicals, instead of commodity chemicals like corn starch or soybean oil. Most of these ideas still are only ideas, but workers are proceeding to put some of the least complicated ones into action. Modification, via genetic engineering, of the ratios of fatty acid components of rapeseed oil is contemplated. Standard breeding to modify chemical components has been used in Canada already, to make rapeseed into a new crop: "canola."

Canola oil, unlike standard rapeseed oil, is essentially free of erucic acid and glucosinolates and therefore is safe for human consumption when used as a cooking oil. Another possibility derives from recent success in cloning the R gene (first identified by Mendel) in peas which provides the opportunity to identify and transfer, *in vitro*, genes giving new starch branching properties and thus, potentially, useful new kinds of specialty starches in old starch-producing crops such as maize or wheat. A word of caution is needed for those looking to replace major portions of commodity crop production areas with specialty-product new versions of the old crops. High value specialty products, by definition, require only small amounts of product to saturate the market. They won't replace very much of the production area now used for production of commodity crops. On the other hand, they will make intensive use of crop breeding and crop management skills. They are "brain-power intensive" crops, and as such, must command higher preunit prices—premiums—or they won't be grown.

## PHOTOLYSIS AND CARBON FIXATION

### Photosynthesis

*Photosynthesis* is the means that primary producers (mostly plants) can obtain energy via light energy. The energy gained from light can be used in various processes mentioned below for the creation of energy that the plant will need to survive and grow. Photosynthesis is a reduction process, where hydrogen is reduced by a coenzyme. This is in contrast to respiration where glucose is oxidised. The process is split into two distinct areas, *photolysis* (the photochemical stage) and the *Calvin Cycle*. The diagram below gives a summary of the reaction, where light energy is used to initiate the reaction in its presence:

$$CO_2 + H_2O \rightarrow \text{glucose} + \text{oxygen}$$

### Photolysis

This part of photosynthesis occurs in the *granum* of *a chloroplast* where light is absorbed by *chlorophyll*; a type of photosynthetic pigment that converts the light to chemical energy. This reacts with water ($H_2O$) and splits the oxygen and hydrogen molecules apart. From this dissection of water, the oxygen is released as a by-product while the reduced hydrogen acceptor makes its way to the second stage of photosynthesis, the Calvin cycle. Overall, since the water is oxidised (hydrogen is removed) and energy is gained in photolysis which is required in the Calvin cycle

### *The Calvin Cycle*

Also known as the carbon fixation stage, this part of the photosynthetic process occurs in the *stroma* of chloroplasts. The carbon made available from breathing in carbon dioxide enters this cycle, which is illustrated below: Just like the *Kreb's Cycle* in respiration, a substrate is manipulated into various carbon compounds to produce energy. In the case of photosynthesis, the following steps occur, which create glucose for respiration from the carbon dioxide introduced into the cycle;

- Carbon from $CO_2$ enters the cycle combining with Ribulose Biphosphate (RuBP)
- A compound formed is unstable and breaks down FROM its 6 carbon nature to a 3 carbon compound called glycerate phosphate (GP)
- Energy is used to break down GP into triose phosphate, while a hydrogen acceptor reduces the compound therefore requiring energy
- Triose Phosphate is the end product of this, a 3 carbon compound which can double up to form glucose, which can be used in respiration.
- The cycle is completed when the leftover GP molecules are met with a carbon acceptor and then turned into RuBP, which is to be joined with the carbon dioxide molecules to re-begin the process.

The energy that is used up in the Calvin cycle is the energy that is made available during photolysis. The glucose that is made via GP can be used in respiration or a building block in forming *starch* and *cellulose*, materials that are commonly in demand in plants.

### *Limiting Factors in Photosynthesis*

Some factors affect the rate of photosynthesis in plants, as follows

- *Temperature* plays a role in affecting the rate of photosynthesis. Enzymes involved in the photosynthetic process are directly affected by the temperature of the organism and its environment
- Light Intensity is also a *limiting factor*, if there is no sunlight, then the photolysis of water cannot occur without the light energy required.
- Carbon Dioxide concentration also plays a factor, due to the supplies of carbon dioxide required in the Calvin cycle stage.

Overall, this is how a plant produces energy which supplies a rich source of glucose for respiration and the building blocks for more complex materials. While animals get their energy FROM food, plants get their energy FROM the sun.

## FUTURE OF AGRICULTURAL BIOTECHNOLOGY

There are divergent views about the rate of uptake of new agricultural biotechnologies and their impact. Kalter and Tauer and the US Office of Technology Assessment anticipate comparatively rapid take-up, while others (Buckwell and Moxey and Farrington do not foresee much impact until the next century, that is until ten to fifteen years have elapsed. The caution of the latter commentators is probably justified, for there are various hurdles to be jumped before biotechnologies can gain acceptance. One hurdle to be overcome is the economic one.

Self-evidently there must be some economic advantage to farmers or agro-industry from adopting the technology. Frequently what captures the scientific imagination turns out to be economically unviable in use, although it is only with use that efficiency can be improved and costs brought down. A more demanding test still for biotechnologies will be to obtain legislative approval and public acceptance. These are intimately connected. In the case of machinery developments, provided safety standards are observed there are no legal impediments to developing new and improved machines, since there are no obvious problems of hazard to the public and therefore public acceptance. Chemical insecticide, fungicides and herbicides do, however, pose much greater problems because of concerns with toxicity, and elabourate testing and licensing procedures have evolved after a relatively haphazard set of procedures in the 1950s and 1960s, when widespread use was made of DDT and organo-phosphorous compounds without adequate recognition of their toxicity.

In significant part, because of the problems which have been experienced with agro-chemicals, the licensing of biotechnologies will inevitably be based on testing at least as stringent as that which now exists for chemicals. Indeed the licensing procedures are likely to be more stringent because of public and scientific concerns. Public acceptance of biotechnology in the food chain will be made more difficult by confusion about differences between biotechnologies. In the early 1980s there was a crescendo of concern about the use of steroids to promote faster liveweight gain in calves and cattle.

The public outcry eventually led to the banning of such hormones for meat animals; but the legacy is a profound distrust of and hostility to new products such as synthetic Bovine Somatrophin (BST), which has the capacity to increase the milk yield of those dairy cows which are 'relatively' deficient in what is a naturally occurring hormone. Although initial scientific results are favourable, and some minor doubts remain, the major influence leading the European Commission to impose a moratorium on the use of BST until the end of 1990 at least is concern about public perception; the Commission has stated 'It would be a serious setback to producers and to the Community's milk policy were present [positive] trends in consumption to be reversed as a result of adverse consumer reaction. With newly bioengineered plants and animals scientific and public concerns are emerging which are of a different order from those associated with previous biotechnology. One relates to the possibility that crop failures may become more frequent if biotechnology leads to a reduction in genetic diversity in crops being grown; such a tendency has already been observed in relation to the Green Revolution technology for wheat. There are also concerns about upsetting the balance of nature in unforseen ways, akin to the unanticipated consequence of introducing rabbits into Australia and then trying to control these by introducing myxomatosis.

Thus questions arise such as what happens if herbicide resistance introduced into a commercial crop plant transfers itself by cross pollination into a closely related weed species, or indeed if the herbicide-resistant crop should colonize wild habitats. There are, of course, more lurid and absurd notions bandied about in the popular press which are similar in nature to the attacks made on Darwinism. The questions, both absurd and real, raised in relation to agricultural biotechnology will undoubtedly slow the rate at which it is adopted. Nevertheless instances of adoption are increasing: BST in the USA, a bioengineered baker's yeast in the UK, bioengineered sheep producing insulin, etc. As these become more widespread and increasingly affect lower-valued, bulk agricultural products, so the impacts of biotechnology on structural change will increase. The balance of economic power will switch increasingly to industry, to high-technology large farms and against smaller farmers in disadvantaged regions and countries. That, unfortunately, is a seemingly inevitable consequence of what we consider to be economic progress.

# 2

# Food Security and Agricultural Biotechnology

Development of sustainable pest management strategies in agriculture has become necessary in view of increasing non-viability of chemical based approach. Among various approaches for the purpose, policy makers have paid far more attention to biotechnological alternatives. The first transgenic variety was approved for commercial trials in cotton.

However, some companies incorporate the Bt gene from the approved varieties in other varieties and released such hybrids to the farmers. It so happened that these hybrids though illegal and released unethically proved more remunerative to the farmers. As if this was not enough, farmers made crosses of this hybrid developed by Nav Bharat Seed Company and developed their own locally suited varieties.

The chapter describes the ethical, business and political dimensions of agriculture biotechnology in India with specific reference to the experience of Bt cotton in Gujarat. The neglect of IPM, herbal pesticides and bio control methods becomes even less justified when state not only tolerates but also encourages widespread experimentation of Bt cotton without any regulation or monitoring. Implications for future policy for technological change have been outlined in the paper.

This chapter deals with the case of Bt cotton in Gujarat, India, as an illustration of how public policy chickens out when large-scale violation of ethical and scientific norms takes place with positive business outcomes.

People have begun to take technological decisions into their own hand. Entrepreneurial spirit has created a unique case of farmer participatory research where on farm crosses are being made between Bt cotton varieties (including the ones which have been released without approval by the regulatory agencies) and other released varieties. New production cycles are being created by farmers who have extended the life of the crop from six months to nine months to reap advantage of continuous flowering and thus higher yield.

All this has happened in a completely unauthorized manner, with full public knowledge and despite complaints of Monsanto and MAYHCO about Nav Bharat Seed Company having 'stolen' their Bt gene. Farmers are happy, politicians do not care and regulatory agencies are satisfied that sending a few committees to enquire is all that they needed to have done.

In a developing country with overflowing food grain stocks, a liberalizing economy set on a growth path, and techno-bureaucracy willing to pursue new technologies with a reasonable open mind, the situation seems ripe for agricultural biotechnology. The investment in this sector is picking up and the private sector has been given a pivotal role along with synergistic public investment.

Top Indian policy makers argue that biotechnology will provide food security to the hungry but do not explain as to why should hunger exist at all in India with its overflowing food stocks, if the issue was only supply! Activists also make a case for biotechnology so long as it increases productivity. The case of Bt cotton in Gujarat shows that this new technology indeed increases productivity, profits of the farmer and eliminates the need for excessive sprays of chemical pesticides.

Yet, when good economic returns follow a bad ethical practice in terms of technological change (not respecting environmental regulation and monitoring, nor intellectual property rights), do ends justify the means? The answer begs for possible alternatives to Bt cotton:

- *The agro-chemical industry and conventional agriculture*: the chemical pesticide industry makes its living so to say, by heavy inputs in cotton (reportedly 40-50 per cent of all the pesticides in India is consumed by hybrid cotton alone). It does not even educate workers and farmers about (a) safe ways of spraying (not almost a single hoarding in the entire country on the subject though there are thousands of bill boards advertising the use of the chemical pesticides), (b) disposing pesticides containers (except in small print in the publications accompanying the pesticide bottles), and (c) practicing integrated pest management – IPM-practices. Ethical violations in chemical pesticide industry are more conspicuous, have hazardous results and are more certain than other new technologies-in fact consequences of excessive use of chemical pesticides on human and environmental conditions are well documented as against the environmental consequences of biotechnology which are still being probed Yet, similar public debate does not ensue.

- *Integrated Pest Management* (IPM): Despite having achieved excellent results through publicly funded IPM, the investment on IPM in Gujarat remained static for the last three years-around 75000 US$ per annum that is less than one percent of the total agricultural budget in Gujarat. This is very likely to be the case at the countrywide level as ascertained in informal dialogues with senior policy makers in the Ministry of Agriculture. IPM offers an alternative to biotechnological solutions in pest control and a cheaper alternative at that. But public policy support for IPM is quite weak.

Even though the adoption of Bt cotton in Gujarat has benefited farmers significantly, the public policy neglect of safer, cheaper and more affordable technological alternatives has obvious ethical implications too. The Department of Biotechnology (DBT) of the Government of India is aware of these alternatives and has made hardly any investment in making even fair comparative trials of Bt cotton with other competing technologies.

Not to mention the almost total absence of investment in public education about the likely hazards of taking up such large scale un-guarded, un-monitored and un-evaluated trials by farmers of the Bt crosses with local hybrids. Undoubtedly, the experience of Gujarat will be recalled in the history of biotechnology as one of the largest trial (with full public knowledge and without any responsible monitoring or evaluation by public agencies at similar scale) of an illegally released technology ever done by people themselves, oblivious of any environmental or other consequence.

## IMPACT OF ECONOMIC REFORMS ON AGRICULTURE IN INDIA

The serious foreign-exchange crisis in 1990 led to a number of well-publicized economic reforms in the early 1990s dealing with trade, industrial licensing, and privatization. The reforms had an impact on the agricultural sector through the central government's effort to withdraw the fertilizer subsidy and place greater emphasis on agricultural exports. The cut in the fertilizer subsidy was a result of the government's commitment to reduce New Delhi's fiscal deficit by removing grants and subsidies from the budget. The government action led to a reduction in the use of chemical fertilizers and protests by farmers and opposition political parties. The government was forced to continue the subsidies but at a somewhat lower level.

New import and export policies aim at enhancing export capabilities of the agricultural sector by increasing productivity and promoting modernization and competitiveness. The measures to facilitate export growth include allowing the import of capital for the agricultural sector, reducing the list of agricultural products that cannot be exported, and removing the minimum export price from a number of products.

Agricultural exports increased by 30 percent in FY 1991 and 14 percent in FY 1992 in terms of rupee value, but declined by 8 percent from FY 1990 to FY 1992 in United States dollar terms because of the devaluation of the rupee in 1991. In the mid-1990s, it was expected that agriculture would continue to be the most important sector of the economy for the rest of the decade in terms of the proportion of GDP.

However, even when it is not the sector providing the largest share of GDP, the importance of agriculture is not likely to diminish because of its critical role in providing food, wage goods, employment, and raw materials to industries.

Despite their preoccupation with industrial development, India's planners and policy makers have had to acknowledge the critical role of agriculture in the early 1990s by changing basic policy. The gains in agricultural production should not lead to complacency, however.

Continuing increases in productivity, developing allied activities in rural areas, and building infrastructure in rural areas are essential if India is to continue to be self-reliant in food and agricultural products and provide a modest surplus for exports.

## AGRICULTURE AND RURAL DEVELOPMENTS

From a nation dependent on food imports to feed its population, India today is not only self—sufficient in grain production, but also has a substantial reserve. The progress made by agriculture in the last four decades has been one of the biggest success stories of free India. Agriculture and allied activities constitute the single largest contributor to the Gross Domestic Product, almost 33% of it. Agriculture is the means of livelihood of about two—thirds of the work force in the country. This increase in agricultural production has been brought about by bringing additional area under cultivation, extension of irrigation facilities, the use of improved high yielding variety of seeds, better techniques evolved through agricultural research, water management, and plant protection through judicious use of fertilizers, pesticides and cropping practices.

## Crops

The 1970s saw a multi-fold increase in wheat production that heralded the Green Revolution. In the next decade rice production rose significantly; in 1995-96, rice production was 79.6 million tons. Total grain production crossed 191 million tons in 1994-95, a big leap from 51 million tons in 1950-5 1. During the Seventh Plan, the average grain production was 155 million tons, 17 million tons more than the Sixth Plan average. To carry improved technologies to farmers, a National Pulse Development Program, covering 13 states, was launched in 1986. The Special Food Production Program augmented efforts to boost pulse production further. In 1995-96, pulse production was 13.2 million tons. With some States offering more than the statutory minimum price, sugarcane production also received a boost, in 1995-96 a record 283.0 million tons was registered.

## Irrigation

As efforts continued to increase the irrigation potential in the country, the last 40 years saw the gross irrigated area reach 8~ million hectares. Flood forecasting has become an important activity over the years. Over 500 hydrological stations collect and transmit data through 400 wireless stations for issuing forecasts for 157 sites. About 5000 forecasts are issued in a year with 94% accuracy. The country also receives international support, with the World Bank as a primary source, for developing the water resources. International cooperation is also envisaged in setting up a National Center for Information on Water and Power. As there is a broad seismic belt in the country, particularly along the Himalayan, the Kutch region and ports of Maharashtra, a scheme is being evolved to collect all data on seismic activity at various dam sites.

## Fertilizers

The fertilizer industry in India has grown tremendously in the last 30 years. The Government is keen to see that fertilizer reaches the farmers in the remote and hilly areas. It has been decided to decontrol the prices, distribution and movement of phosphatic and potassic fertilizers. Steps have been taken to ensure an increase in the supply of non-chemical fertilizers at reasonable prices. There are 53 fertilizer quality control laboratories in the country. Since bio—fertilisers are regarded as an effective, cheap and renewable supplement to chemical fertilizers, the Government is implementing a National Project on Development and Use of Bio-fertilisers. Under this scheme, one national and six regional centers for organizing training, demonstrating programs and quality testing of bio-fertilisers has been taken up. It was a challenging decision of the Government to take Bombay High gas through a 1,700-km pipeline to feed fertilizer plants located in the consumption centers of North India. However, the major policy which has ensured the growth of the fertilizer industry is the thrust on accelerating fertilizer consumption by fixing, on the one hand, low and uniform price for fertilizers, and on the other hand providing the manufacturers adequate compensation through the retention price and subsidy scheme. As expected, fertilizer nutrient demand has gone up from 0.29 million tons in 1960-61 to 13.9 million tons at the end of 1995-96, compared to 12.15MT during 1992-93.

## Fisheries

Fish production achieved an all-time high of 4.9 million tons at the end of 1995-96. Programs that have helped boost production include the National Program of Developing Fish Seeds, Fish Farmers' Development Agencies and Brackish Water Fish Farmers' Development Agencies. The Central Institute of Fisheries Nautical and Engineering Training trains the necessary manpower. To diversify fishing methods and introduce processed fish products on a semi-commercial scale, an Integrated Fisheries Project has been launched. A National Fisheries Advisory Board has also been established.

## Food Processing

A Ministry of Food Processing Industries was established in July 1988 to ensure better utilization of farmers' output by inducting modem technology into the processing of food products, thus augmenting the income of farmers and generating employment opportunities in rural areas. A new seeds policy has been adopted to provide access to high quality seeds and plant material for vegetables, fruit, flowers, oil-seeds and pulses, without in any way compromising quarantine conditions. Initiatives have been taken to encourage private sector investment in the food processing industry.

## Agricultural Research

The apex body for education, research and extension education in the field of agriculture is the Indian Council of Agricultural Research (ICAR), established in 1929. India's transformation from a food deficit to a food surplus country is largely due to ICAR's smooth and rapid transfer of farm technology from the laboratory to the land. ICAR discharges its responsibilities through 43 research institutes, four national research bureaus, 20 national research centers, nine project directorates, 70 all-India coordinated research projects, and 109 Krishi Vigyan Kendras (farm science centers). Besides, the program of Agricultural Education is coordinated by ICAR with the curricula and other normative guidance given to the 26 Agricultural Universities and four National Research Institutes.

### Oilseeds Production

A Technology Mission on Oilseeds was launched in 1986 to increase production of oilseeds in the country and attain self-sufficiency. Pulses were brought under the Technology Mission in 1990. Before the Mission was launched in 1985-86, oilseed production was 10.83 million tons; during 1995-96, it was estimated at 22.42 million tons, which is a record. Soybean, rapeseed and mustard largely contributed the increase in production. Production of pulses has seen many ups and downs, which is expected to be checked under the Mission. The country grows mainly nine oilseeds, with groundnut, rapeseed and mustard accounting for 62% of total production. Lately, soybean and sunflower have shown major growth potential.

### Drinking Water

A Technology Mission on Drinking Water and Related Water Management has been constituted to cover the residual problem villages and provide potable water at 40 liter per capita per day, and 70 liters per capita per day in desert areas inclusive of 30 liters for cattle. The Mission is tackling the problem through 55 mini-missions in project districts and countrywide problem oriented sub-missions. A Village Level Operation and Maintenance (VLOM) pump called India Mark-11 has been developed and is being exported to 40 countries. By March 31, 1993, over 79% of the rural and about 85% of the urban population was provided drinking water facilities.

## THE FIELD SURVEY

### Survey Outline and Methodology

A sample of 363 farmers from various parts of Gujarat except Kutchh was surveyed to collect data about the experience with Bt cotton during 2001-2002, the year when Navbharat 151 seed was formally sold by the Navbharat seeds ltd. During last two years, the seed has been not sold by this company.

But the seed has been multiplied by the farmers or other private agencies as mentioned above. The sample survey has been organized on the basis of source of seed in terms of a location and producer.

The respondents were selected at random from 75 villages from 10 major cotton-growing districts of Gujarat state. The villages were selected on the basis of scatter of the technology to capture as much agro-ecological diversity as possible.

The respondents were interview by the Post Graduate students of Gujarat Agricultural University pursuing their Masters Degree in Plant Breeding and Genetics, Agril. Entomology, Agronomy and Agril. Economics. The respondents were visited at their farms and the pre designed and tested questionnaires were used to elicit their opinions.

Since the survey is based on recall data and that too with a lag of one year, there is a possibility of some loss of information. The data pertains to the year 2002. The information collected from the respondent farmers was crosschecked by asking the neighboring farmers about their views relating to the BT cotton crop of the respondent farmer. Several farmers were hesitant in giving their names since they knew that their cultivation of Bt cotton was illegal. Once assured of anonymity, they agreed to cooperate in the survey.

## Field Survey Results

We provide brief information about the sample of farmers who bought seed from distant locations, local shop, other sources with in their region and those who have made it on their own. There was not much difference among these groups in regard to age composition or education.

The largest proportion of farmers in all land size groups bought seeds from distant sources signifying a distinct preference for the BT seeds even from far off places. The main difference was that small and large farmers had sizeable section which purchased seeds from shops (which had highest productivity and were thus perhaps of best quality).

The key contrast was that while majority of marginal land holders had only less than one fourth area under irrigation, compared to those who have bought from shop and from distant places having much larger area under irrigation. Among those who has procured seeds from shops and distant places, majority had irrigated holding signifying their ability ti provide better production environment to good quality seeds.

The sample characteristics are on the basis of seed producers (companies or farmers themselves). The age and education did not make much difference to seed buying behavior. However, if we look at the ratio of branded to non branded seed purchase behavior, the ratio was 3:2 among marginal land holders while it was about 2:1 in the larger land holding farmers.

If the land holding is a proxy for better economic power, those who oculd afford, did go for better seeds. What is interesting though is that even among these more opportunistic buyers, a significantly higher proportion went for illegal Navbharat seeds rather than Monsanto's legally released Bt seeds.

Poorer farmers relied much more on low quality $F_2$ seeds compared to the rest. A discriminant analysis among the four groups also revealed the average yield to be lower among growers of own seed or $F_2$ seed by about 10 to 25 per cent.

Compared to normal cotton both MMB and Navbharat flowered earlier, took lesser days to maturity, was slightly dwarf and had higher number of balls per plant. The trend was similar in the case of farmers' own seeds as well as the $F_2$.

The average yield was highest in the case of MMB followed by Navbharat, farmers crossed and $F_2$. When one looks at the source in terms of location the seeds obtained from shops apparently performed the best followed by the ones bought from distant sources, from other farmers, and the minimum yield was from own seed. The experience with different pesticides was on an average better in the case of Navbharat seed than in case of MMB or other sources. When one looks at the composition of sources, location and producerwise, among all the producers, the seed of Navbharat was obtained by largest number of people from far off places indicating maximum demand and minimum availability from the shops. In terms of farmers crossed $F_2$, pattern was similar except that shop was not an important source of farmers crossed or $F_2$ seed. There was not much difference in terms of the incidence of different pests. The survey by and large reveals that farmers growing MMB have achieved marginally higher yield but at higher costs ( since as mentioned earlier, MMB seeds costed about 35 usd per 450 gms, where as Navbharat seeds bought from shop costed about USD 9-10). Whereas in case of Navbharat seeds or its derivatives, they have still managed much higher yields than the normal hybrids at much lesser cost.

From the risk perspective, what is most remarkable is the higher standard deviation (SD) in case of Bt cotton compared to the normal hybrids. Higher risk and higher return would make the technological choice attractive for entrepreneurial farmers but for the others, it might be inhibitory. Given the fact that the survey is based on recall data and that too with a lag of one year, there is a possibility of some loss of information. We are planning to repeat this survey among the same farmers for the current year's performance. What has been demonstrated unambiguously is that in the current year a very significant number of farmers have decided to irrigate the crop even after normal season and elongate the life by about three months with very attractive returns. Normally farmers grow potato in some parts after cotton but this year they decided to continue with the cotton, particularly the Navbharat seeds and its derivatives.

## AGRICULTURE IN INDIA

Agriculture in India is one of the most prominent sectors in its economy. Agriculture and allied sectors like forestry, logging and fishing accounted for 18.6% of the GDP in 2005 and employed 60% of the country's population. It accounts for 8.56% of India's exports. About 43% of India's geographical area is used for agricultural activity.

Despite a steady decline of its share in the GDP, agriculture is still the largest economic sector and plays a significant role in the overall socio-economic development of India. The monsoons play a critical role in the Indian sub-continent's agriculture in determining whether the harvest will be bountiful, average, or poor in any given year. The entire rainfall in the sub-continent is concentrated in the few monsoon months.

Agriculture in India is constitutionally the responsibility of the states rather than the central government. The central government's role is in formulating policy and providing financial resources for agriculture.

The government administers prices of essential commodities to protect farmer's interests. It also administers other commodities which are produced by government-controlled companies, like petroleum, coal, nitrogenous fertilizers, etc. Other than these, most agricultural commodity markets operate under the normal forces of demand and supply.

## HISTORY OF INDIAN AGRICULTURE

Evidence of the presence of wheat and some legumes in the 6th millennium BC have been found in the Indus Valley. Oranges were cultivated in the same millennium. The crops grown in the valley around 4000 BC were typically wheat, peas, sesame seed, barley, dates and mangoes. By 3500 BC cotton growing and cotton textiles were quite advanced in the valley.

By 3000 BC farming of rice had started. Other monsoon crops of importance of the time was cane sugar. By 2500 BC, rice was an important component of the staple diet in Mohenjodaro near the Arabian Sea. The Indus Plain had rich alluvial deposits which came down the Indus River in annual floods. This helped sustain farming that formed basis of the Indus Valley Civilization at Harappa. The people built dams and drainage systems for the crops. By 2000 BC tea, bananas and apples were being cultivated in India. There was coconut trade with East Africa in 200 AD. By 500 AD, egg plants were being cultivated.

## BIOTECHNOLOGY CENTER IN NE INDIA

The NE region is rich in diversity-both culturally and agriculturally. With agriculture being the predominant industry of the region, any positive changes focused on here is welcome news for 70% of the population. Since land holding is limited, any increase in agriculture product must come from either changing the cultural practices or integration of agricultural biotechnology. The later is attractive and promising for the future since there is a limit to what cultural practices will add to the overall production. For example, the Green Revolution in the early 1960s with the introduction of dwarf varieties of wheat and rice depended on cultural practices to make the changes, but production leveled off after agricultural inputs where enacted.

Establishing a biotechnology center is imperative if the region is to improve the agriculture to meet the future food demands of the region. Crop improvement through genetic engineering must be a priority if the region wants to stay competitive and to best utilize natural resources. A number of states have already established their own biotechnology center as is in the state of Andhra Pradesh. The region already has a number of universities that can provide the technical know-how and skills needed in the hundreds of competent faculty and scientists they employ. Additionally, there are several Indian Council of Agricultural Research centers in the region, which may have already thought through the process of establishing a biotechnology center.

For a biotechnology center to stay competitive, the center must reflect the following criteria:

(1) as a government (central and regional) sponsor with a non-political entity,

(2) governed by a board of directors (scientist should make up a majority of the board),

(3) registered as a non-profit organization,

(4) development process driven by product delivery rather than consumption driven,

(5) seek its own funding through government and foundation, both national and international,

(6) competitive funding scheme should ensure the integration of biotechnology with classical breeding and other associated technical disciplines,

(7) located close in proximity to a major airport,

(8) close to a major university or college,

(9) have the goal of becoming a self-supporting entity,
(10) have the objective of addressing regional issues,
(11) employee selection criteria strictly merit, and
(12) the seven sister states each contribute and play a role.

## WHAT ARE THE ECONOMIC IMPACTS AND THE MARKET OUTLOOK FOR THE REGION WITH GMO?

The economic impact on the region is enormous-from food import dependency to a player in India and the world food market. The total area under GM crops globally has risen by 20 per cent in the last year alone. In India, where GM use was allowed only three years ago, about 1.34 million hectares have come under B.t. cotton. A cotton farmer in India can increase his bottom line even though he pays higher seed premium in order to grow genetically engineered cotton. The market outlook for GM products in India is excellent since food shortage and poverty are still alive in India.

In conclusion, I believe biotechnology can bring unprecedented economic impact in the region, which in turn will change the attitude of the people. I believe if a product and market driven economy takes a hold in the region, the young people will trade their guns for the newfound economic power they so desperately seek to achieve. Rather than stealing at gunpoint, they will work hard exploiting the new economic reality within their reach.

One final note, but economic freedom is only one of three freedoms I mention in an earlier article, True Freedom.? One cannot be truly free if he is not free spiritually. Spiritual freedom includes freedom from corruption, murder, envy, and all kinds of evil. Poverty and hunger in India can be eliminated, but the greater challenge is to eradicate the spiritual poverty that is beneath the surface of each person's apparent happiness.

## IPRS AND INVESTMENTS IN PLANT BIOTECHNOLOGY

The implicit rationale for IPR protection is that they promote investments in plant breeding and bio-engineering. This hypothesis is contested by section of analysts. Nevertheless it is a fact that the structure of IPRs do influence investments. IPRs induce their own pattern of innovations. The time taken for R&D to fructify as innovations is a crucial aspect guiding investments.

Often IPR regimes play a key role in influencing the time pattern of innovations. It is a well-known fact that conventional plant breeding methods are dilatory and time consuming. It requires 7 to 8 generations of repeated breeding to reduce heterozygosity of new genotypes. As Vossen (1985) observes

Arabica Coffee subject to three to four cycles of breeding still display heterozygosity and it is not until 20 to 25 years (spanning 4 to 5 generations) that one arrives at true breeding varieties. This time lag is even more pronounced for the Arabusta coffee variety (inter-specific hybrid of Arabica and Robusta Coffee) where segregation of unfavourable characteristics could still occur even after several generations of backcrossing and selfing.

Another interesting example is the one cited by Evenson (1991) about the long gestation period in regard to rice varieties in India. According to Evenson, India released 306 rice varieties for planting during the period 1965-1986 after performing 20,000 crosses over a period of over 15 years since 1950.

It must be understood that this long process of backcrossing was preceded by considerable time spent on location and identification of individuals plants with requisite traits. The speed or pace at which plant varieties qualify for being released for planting have economic implications on investment behaviour in seed/plant breeding industry. Asexual modes of plant breeding which include plant genetic 'engineering' and transgenic techniques could obviate the long gestation period associated with pre-release preparations encountered conventional breeding.

It is a different matter than biosafety concerns may delay releases of these varieties. The time lag in release of plant varieties fulfilling the criteria of novelty, distinctness, uniformity and stability (or alternatively the requirements of novelty, inventiveness and commercial application) raise fundamental issues from the view-point of capital investments. The time lag in varietal release is potentially least in the case of transgenic and non-sexually produced varieties while for traditional or classical methods the time lag can be greater. The UPOV 1991 based plant IPR regimes could induce longer time lags as compared to the situation when an IPR regime is based on the 1978 version of the UPOV. This is due to the fact that the minimum genetic distance concept implicit in UPOV 1991 precludes 'close cousin' varieties from acquiring protection rights, unless the techniques of genetic distance determination by themselves are inadequate.

UPOV 1978 would facilitate capital investments to some extent as the release of plant varieties is periodically more frequent. Even under the plant patent regimes, investments could bring about short-term returns, though the volume of investments will have to be larger on account of capital intensity of advanced plant genetic engineering technologies.

By comparison, UPOV 1991 regime is least conducive to capital investments as plant variety releases can be painfully slow under this regime.

Therefore sustenance of conventional plant breeding under UPOV 1991 regimes, can only be ensured by infusion of low discount capital.

Paradoxically it is the reduced availability of low discount public funds for plant breeding that has induced countries like India to go in for Plant Varieties Protection Legislations modelled on UPOV lines. The implicit assumption in undertaking this step is that private funds for R&D in plant breeding will 'crowd in', if protection plant varieties rights are accorded. But the reality is that in liberalised deregulated developing economies, private investments have 'crowded in', in short return promising avenues.

In other words, private investment funds for R&D are not likely to be channellised into medium to long return yielding areas. Let ra, rb & rc be the discount rates relevant for regimes A, B & C in Figure I. Then the relative rates of discount will be r1 > rb < rc and relative present worth will be pa < pb > pc.

Consequently classical plant breeding activities will suffer for want of funds in the UPOV 1991. In other words, UPOV 1991 modelled national legislations will not necessarily hold attraction for private sector investors in R&D projects involving classical plant breeding as compared to say, UPOV 1978 or the Plant Patent regimes. Rather private investments would be catalysed better by *sui-generis* plant protection legislations, which partake of plant patent elements and focus in the main on hybrids and genetically engineered varieties.

Given the linkages between plant breeding and the seed industry these perverse investment behaviour trends can produce major economic consequences. The Seed Laws in various countries, both advanced and developing, are getting to be progressively oriented towards restricting market circulation to varieties that are conferred with plant breeder or patent rights. The UPOV guidelines on DUS are used world wide not only as the basis for establishing varietal distinctiveness and descriptions but also for seed certification purposes. The DUS guidelines are employed for recognising and registering not only the 'basic' seeds used for multiplication of 'pedigree' seeds but also for certification of the pedigree seeds themselves.

Evenson (1991) explains how absence of IPR rights will reduce Marginal Variable Costs (MVC) of seed multiplication. Tougher controls by certification of seeds could ensure that seed supply is kept in a 'constricted mode' and MVC of seed multiplication are raised. This in turn, will produce its own supply-demand dynamics as *Figure I* depicts. Figure I brings out the implications of IPR regimes in terms of the supply curves for seeds. As Figure I shows that in a situation of 'liberal' plant varieties protection laws (which do not provide for genetic distance) and equally liberal 'seed laws' which permit non-registered varieties (including landraces) to be sold in open markets, the macro-supply curve of seeds will be elastic (S1). When these regimes shift towards 'genetic dependency' PBR regimes and tighter seed laws that prohibit marketing of non-registered varieties, the supply curve will tend to be relatively inelastic (S2).

Analogously, demand curves could shift from D1 (wherein demand for protected varieties will be elastic on account of availability of landraces and non-protected varieties in the markets) to D2 (where the demand for the protected variety will be inelastic on account of ban on sale of landraces and other non-protected varieties).

In the D1-S1 situation, supply constraints and demand flexibility could induce low price regimes while in D2-S2 the reverse situation would occur. D2-S2 situation is not desirable for countries like India for other reasons as well.

The experience of transitional economies of East Europe, which had re-modelled their Seed Laws in conformity with European Community Seed Laws, has been sad. Farmers of the East European bloc cultivating and conserving landraces have been adversely affected by new seed laws, which by prohibiting trade in non-registered varieties not fulfilling the DUS criteria have contributed to the erosion of landraces in these countries. It is thus clear that agro-biodiversity can be adversely affected by the uniformity criteria of DUS, which brings about resultant changes in demand and supply position of seeds.

## POST R&D REGULATIONS AND INVESTMENTS

Post-R&D regulations play a critical role in deciding on the pace of investments in the biotechnology sector. Even if a start-up enterprise enjoys a favourable IPR regime, that rewards its inventions, it is unlikely that it will invest in product development, in case post-R&D regulations are rigid and dilatory in nature. This then forms the backdrop for strategic alliances between start-up ventures and downstream industrial complexes. Multi-layered regulations in the post-R&D phase involve complexities which are expressed in the delays in commercial application /utilization of invented products or technologies. In the case of third generation plant biotechnologies, such as transgenics, the problem is compounded by the fact that regulations by themselves are complex and less understood by the regulators themselves. A transgenic plant would have to undergo multi-stage regulatory checks and clearances before finding commercial application.

India is uniquely situated as far as biotechnology investments and infrastructure are concerned. In the area of plant biotechnology, advances have been made in transgenics of rice and wheat carrying stress tolerant genes such as CodA, COR47 and HVA1. The larger challenge is to take these products through the regulatory processes. More formidable is the issue of market development and appropriate pricing of transgenic seeds. There are many imponderables here. Market shares for seeds in India have varied from crop to crop and from region to region. The 'demand' for transgenic plants or seeds may be even more uncertain, given the 'safety' dimensions and risk aversion propensity of Indian farmers.

In the dry agricultural pockets of Central and South India, farmers tend to be attached to traditional drought-hardy varieties of plants, which in turn contributes to their initial 'inertia' to shift to transgenics. Thus the task of marketing and pricing of novel biotechnology plants gets to be complex in such environment of 'demand incoherence'. A high margin pricing strategy for seeds may choke demand for transgenics.

Thus the degree of risks in commercializing transgenic plants could be of a high order. Given these realities, the idea of strategic alliances between start-ups and downstream entities looks to be realistic. The fact that premier R&D institutions such as the Indian Institute of Science, has set up its own start-up biotechnology venture, renders a strategy of tie-ups, a logical option. Much will depend upon the complexities inherent in product development and the willingness of competent downstream entities to signup agreements. To a large extent, this will depend upon enabling policies at the upper end of the value chain.

To sum up, India's agricultural biotechnology industry has the necessary policy and regulatory support to promote active R&D. However, in the critical area of product development and commercialization much more attention needs to be paid. The multi-tiered regulatory framework that exists could seriously affect the performance of start-up companies that would like to run through the entire life-cycle of biotechnology product development. Consequently, partnerships and strategic alliances should be encouraged between start-up ventures and established companies that have high product development capabilities. The experience of other countries in the Asia Pacific region could be a key input in framing development strategies for biotechnology products in India.

# 3

# Agricultural and Crop Biotechnology

Since time began we humans have always relied on plants and animals to provide us with food, shelter, clothing and fuel, and for thousands of years farmers have strived to better these in order to continue providing us and to meet the needs of evolution. As the population grows and expands then so does the need for resources provided by plants and animals.

With the population expected to reach over 10 billion by the year 2030 it is estimated that the world food production will have to double on the farmland that exists today, if it is to compete with the anticipated growth in the population. The ever-increasing needs of yields of crops and the decreasing crop inputs such as water and fertiliser can all be helped by biotechnology as can advancements in providing pest control that is better for our environment.

## CROP BIOTECHNOLOGY

Horticulturists and farmers have always relied for centuries on cross breeding and hybridizing plants and making genetic modifications to them in order to make improvements in the output and quality of food and fibre crops. They have made great advancements in building in protection against pests and diseases and protection from the harsher elements.

We learned from the Stone Age farmers to make a selection for next years crops from those plants that were strongest and healthiest and to save seed from these with which to replant. By selectively sowing seeds the earliest agriculturists performed modifications genetically to turn plants growing in the wild into domesticated crops, a long time before the science of genetics was really understood.

Over the years, knowledge and understanding of plant genetics improved and we gradually began to purposefully cross breed plants with the traits we wanted in order to produce plants which had the benefits of both parents, leaving undesirable traits behind. Virtually every plant that is grown for food today has come from cross breeding or hybridization or a combination of both.

Although we have come a long way in such a short time the process we were using to produce these crops was time consuming, costly and inefficient. Here is where the tools of biotechnology come into play, biotechnology allows us to take single genes from plants that give us the desired traits and move them freely from one plant to another.

This is a far more precise process than what we used to use and eliminates the thousands of genes of unknown function which was transferred along with the good traits into our crops. Biotechnology can also help us to remove the technical obstacles when moving genetic traits between plants and opens up a whole new world in benefiting food production.

## HOW BIOTECHNOLOGY CAN IMPROVE CROP PRODUCTION

Agricultural scientists are incorporating biotechnology into crop production and protection, using the same traits that have been used throughout the decades and incorporate cross breeding and genetic modifications. Techniques are being used to increase the size of the yield of the crops, make crops resistant to disease caused by bacteria, fungi and viruses and give them the ability to better withstand harsher elements.

Biotechnology has also given us a better understanding of how we can make better use of the plants built in defence systems and it has opened up new avenues allowing us to work with nature by providing us with new bio pesticides.

These new pesticides allow us target crop pests but do not harm humans, animals, birds, fish and insects; they are also beneficial as they can control pests which have become resistant to the more conventional pesticides.

## AGRICULTURE IN THE CHANGING GLOBAL SCENARIO

Steady globalization of trade has profound implications for future agricultural development. The diversity of India's agro-ecological setting, high bio-diversity and relatively low cost of labour provide potential for agricultural competitiveness in a globalized economy. It is expected that with increasing globalization of markets over the years there will be demands for agricultural intensification. This will also be favoured because of greater backward and forward linkages between agriculture and food industry.

Therefore, increase in production and productivity are bound to be strategically important to economy. Intensification will not only favour alleviation of rural poverty but will also improve resource conservation particularly in the small farming sector where farmers can be encouraged to take up organized production of high value crops such as fruits, specialty vegetables, flowers medicinal and aromatic herbs etc. Stronger demands for crops of the small farmers' will not only improve incomes and welfare but will also make investments in technology and resource conservation more attractive. The General Agreement on Tariff and Trade (GATT) and liberalization of global trade is bound to have impact on future land use and production pattern. Understanding the local, national and international environment under which agricultural production is taking shape will be crucial in developing our own strategies.

## EXTENSION STRATEGIES

Since early fifties a number of public by funded agricultural development programmes have been sponsored. These have included programmes like the National Extension Service (NES) Blocks in 1953, the Intensive Agricultural District Programme (IADP) in 1961-62, the Intensive Agricultural Area Programme (IAAP) 1964-65, the High Yielding Variety (HYV) programme 1966-67 and the Small and Marginal Farmers' Development Programmes (SMFDP) in 1969-70.

Though these programmes had a perceptible impact the efforts did not get replicated over different areas and categories of farmers. In mid seventies based on pilot level project in Rajasthan Canal and Chambal command area a 'Training and Visit' (T&V) system of extension was promoted in different states.

Extension efforts of the Indian Council of Agricultural Research through its research Institutes and the State Agricultural University were largely limited to demonstration of new technologies through such programmes as National Demonstration Project, Operational Research Project, the Lab to Land Programme and the Krishi Vigyan Kendras. However, there appears much to be desired in the way that extension programmes are conceived and implemented.

At present extension programmes are implemented in largely a top-down fashion leaving little scope for localized planning and action. Farmers are almost passive receivers and their involvement in the process of technology generation and adoption is almost absent. Extension services, at present, are almost exclusively in the public sector domain and there is no effort or institutional support for other operators e.g. the NGOs, the corporate bodies etc. Extension programmes sponsored by the government operate largely in isolation and there appears a strong need to view the extension programmes as an integral part of the research and development process.

The challenges facing agricultural development call for fundamental changes in our approach to technology transfer/extension programmes. Changes are necessary in the context of changing economic environment following policy adjustments in relation to privatization, deregulation and globalization calling for greater efficiency and effectiveness of the extension system. More importantly there is need for

- Greater emphasis on providing producers with knowledge and understanding needed to overcome the problems or to exploit opportunities of their own specific production systems. Correspondingly there will be a need to de-emphasize 'package of practices' or the blanket recommendations, top down approach followed thus far.
- Shift in the focus of public extension systems from promoting inputs use to one on sustainable management of resources and improvements in the production system as a whole.
- Closer interaction between farmers, extension scientists and production system researchers in diagnosing problems and identifying location specific recommendations emphasizing participation and education rather than being prescriptive.
- Widening the range of extension delivering agencies. While the publicly operated extension systems will continue to be important, there will appear a greater role for NGOs, farmers' associations and corporate sectors in particular situations. Role of commercial suppliers of seeds, agrochemicals, machinery, vaccines and medicines in providing advisories, as is already being done in a limited way, will need to be encouraged and factored into public system's own priorities.
- Wider and more creative use of mass media in tune with current developments in information technology to get information across to the farming community whose ability to overcome constrains at farm level will increasingly depend on access to reliable and up-to-date information.

## CLASS ANALYSIS OF INDIAN AGRICULTURE

No doubt, a growth in "consumption" and "distribution" (i.e. growth in the home market) is an indication of the growth of capitalist relations in production but the important question is that what is its impact on production relations, and, even with this growth of commodity production which of the two modes of production (i.e., capitalist and feudal) is predominant in Indian agriculture ? This can only be done by taking an overall picture of production relations and not merely by picturing the growth in "consumption" and "distribution".

Lenin has said, *"that the economists who have discoursed at length on the inadequate attention paid by the classical economists to 'distribution' and 'consumption' have not been able to give the slightest explanation of the most fundamental problems of 'distribution' and 'consumption'. That is understandable, for one cannot even discuss 'consumption' unless one understands the process of the reproduction of the total social capital and of the replacement of the various parts of the social product... It is not with 'production' that political economy deals, but with the social relations of men in production, with the social system of production. Once these social relations have been analysed, the place in production of every class, and consequently, the share they get of the national consumption is thereby defined."* (Development of Capitalism in Russia........Lenin)

Capitalism does not merely entail commodity production. As Lenin has said, *"Capitalism is commodity production at its highest stage of development, when labour power itself becomes a commodity."* (Imperialism the Highest Stage of Capitalism.... Lenin) Besides, when the CRC, Communist League, etc., speak of increasing "consumption" and "distribution" they forget Marx's basic theory of realisation which states that:

*"According to the general law of capitalist production, constant capital grows faster than variable capital"* and that *"capitalist production, and, consequently, the home market, grow not so much on account of articles of consumption as on account of the means of production. In other words, the increase in means of production out-strips the increase in articles of consumption."* (Development of Capitalism in Russia.......Lenin)

In other words, while dealing with the growth of commodity production and capitalist relations in India, there is a need to picture the nature of the growth-in which spheres is it confined to, and to what extent is it confined to consumer items and to what extent capital items. Is it the sphere of productive consumption that is increasing in agriculture or is it the sphere of consumer expenditure that has increased ?

While analysing the mode of production in Indian agriculture it is necessary to do so based on the laws of capitalist production as discovered by Marx and further developed by Lenin. These scientific laws of capitalism and imperialism are not time bound. They neither get outdated nor change from country to country. The conditions may change in time and space but the basic laws of capitalism and imperialism remain unchanged. To understand the mode of production in India it is no use negating these laws, as the CRC does, but utilise them to understand the mode of production in India.

Regarding the question of capitalism in agriculture an important principle outlined by Lenin was that, *"it is the development of capitalism in the manufacturing industry that is the main force which gives rise to, and develops, capitalism in agriculture."*

Also, while talking of the transformation of pre-capitalist forms of production to capitalism, Lenin has said that:

*"the socialisation of labour by capitalism is manifested in the following process:*

*Firstly, the very growth of commodity production destroys the scattered condition of small economic units that is characteristic of natural economy and draws together the small local markets into an enormous national (and then world) market. Production of oneself is transferred into production for the whole of society: and the greater the development of capitalism, the stronger becomes the contradiction between this collective character of production and individual character of appropriation.*

*Secondly, capitalism replaces the former scattered production by an unprecedented4d concentration both in agriculture and in industry.*

*Thirdly, capitalism eliminates the forms of personal dependence that constituted on inalienable component of preceding systems of economy...*

*Fourthly, capitalism necessarily creates mobility of the population, something not required by previous systems of social economy and impossible under them on anything like a large scale.*

*Fifthly, capitalism constantly reduces the proportion of the population engaged in agriculture and increases the number of large industrial centres.*

*Sixthly, capitalist society increases the population's need for association, for organisation, and lends these organisations a character distinct from those of former times. While breaking down the narrow, local social-estate associations of medieval society and creating fierce competition, capitalism at the same time splits the whole society into large groups of persons occupying different positions in production and gives a tremendous impetus to organisation within each such group.*

*Seventhly, all the above mentioned changes effected in the old economic system by capitalism inevitably lead also to a change in the mentality of the population. The spasmodic character of economic development, the rapid transformation of the methods of production and the enormous concentration of production, the disappearance of all forms of personal dependence and patriar-chalism in relationships, the mobility of population, the influence of the big industrial centres, etc., all this cannot but lead to a profound change in the very character of the producers."* (Development of Capitalism in Russia, by VI Lenin) With this brief introduction on the approach to the problem, let us now turn to study the relations of production in Indian agriculture.

While studying the relations of production in Indian agriculture we will examine it under two heads: a) Growth of Commodity Production, and b) Disintegration of the peasantry.

## GROWTH OF COMMODITY PRODUCTION

The growth of commodity production in agriculture can be assessed by the extent of utilisation of inputs and the growth of outputs produced for the market. The 'green revolution', which was introduced by the World Bank in Third World countries in the mid-1960s, was part of the imperialist policy to penetrate the countryside for markets.

This 'green revolution' has led to certain changes in agriculture which must be analysed. This section will be classified under the following sub heads: 1) 'Green Revolution', 2) Inputs, 3) Credit, 4) Output and Productivity Trends, and 5) Utilisation of Surplus.

### *1) Green Revolution*

### *2) Inputs*

Though irrigation may have increased, out of the 350 million acres of net cropped area, barely 50 million acres, or 14%, is cultivated more than once a year (VI Five Year Plan figures). If one crop takes on an average four months, 86% of India's agricultural land remains idle for 66% of the time per year. To put it differently, nearly 60% of the cropped area is not at all used. If agriculture was to be put on a capitalist footing and capitalist farming was to predominate in India, there would not be such a gigantic wastage of land utilisation and such a slow growth of irrigation facilities.

But besides the modernisaation, the number of wooden ploughs increased from 37.5 million in 1956 to 43 million in 1966 and further to 44.5 million (i.e. during the green revolution period) in 1972

### *3. Credit*

It is thus clear that the stranglehold of usurious capital still dominates all sections of the rural populace. But usurious is an indication of pre-capitalist, feudai relations of production, which sucks up the surplus and prevents it from seeking productive channels. Besides, this usurious capital is increasing day by day. Marx has said that, *"Usurer's capital as the characteristic form of interest bearing capital corresponds to the predominance of small scale production of the self-employed peasant and small master craftsman", "Usury centralises money wealth where the means of production are dispersed. It does not alter the mode of production, but attaches itself firmly to it like a parasite and makes it wretched. It sucks out its blood, enervates it and compels reproduction to proceed under even more pitiable conditions."* (Karl Marx, Vol. III, Chapter 36, pp594)

.............Also with a large section of the peasantry having turned to HYV, (and now unable to turn back to the traditional varieties) and caught in a crisis with diminishing returns for their output, the peasants are dependent on large capital inputs each year which they are now unable to finance. This is forcing them to seek larger and larger loans. With the peasantry unable to even pay their interest on their cooperative loans, they are turning more and more back to the moneylender. This trend is bound to increase enormously in the coming years as the crisis in agriculture deepens.

***4. Output and Productivity trends***

In fact, the annual rate of growth of production of all agricultural crops actually dropped in the second period; while it was 3.2% in the 1951-52 to 1964-65 period, it was just 2.6% in the 1964-65 to 1983-84 period. This was because the increase in area under crops increased phenomenally in the first period at an annual rate of 1.7%, while in the second period the annual in-crease was just 0.4%. In other words, much better results in agricultural production could have been achieved by increasing the area under cropping than by introducing HYV. The economic and scientific research foundation has estimated that soil erosion and deforestation has been so acute that India looses about 1% of its cultivable land every year to deserts. It is said that out of a total of 306 million hectares of cultivable land, 145 million hectares are either threatened with erosion or badly in need of soil and water conservation measures.

Generally, it can be summed up that production and yield of cropping has increased, with small increase towards cash crops, indicating some growth in capitalist relations. This is indicated through the change in some pockets of agriculture to capitalist farming and growth in commodity relations in large tracts of semi-feudal production, which continue their semi-feudal existence with enhanced contradictions caused by this lopsided growth of commodity relations within it. Productivity trends continue to indicate a backward mode of production with large farms being the least productive.

Also with the increase in value of output being nowhere commensurate with the increase in value of inputs, it is clear that the growth in commodity production is lopsided, with this growth merely facilitating the dumping of commodities of the imperialists and comprador big bourgeoisie into the rural sector without significant returns to the farmer. This has led to a new set of contradictions in agriculture which are bound to intensify.

*5) Utilisation of Surplus*

This, in fact, is one of the factors in the determination of capitalist growth, as a fundamental law of capitalism is that constant capital (i.e., production of means of production) must grow at an increasingly fast rate. Therefore, in capitalist farming the kulak must reinvest his surplus in the farm in improved technology. Thereby he would increase the productivity of his farm and the surplus value extracted from it. Besides this low generation of surplus in agriculture, even that generated goes primarily to other spheres. With the return on moneylending and trading far more profitable, a large proportion of this surplus is not reinvested in agriculture, instead finds its way in such spheres of activity. In fact, as long as usury continues to dominate the countryside as a most profitable sphere of investment, it will restrict the growth of capitalist develop-ment. Of course, with the inception of the 'green revolution' things have somewhat changed; but the limited extent to which it has generated capitalist farming can be seen from the poor levels of capital investment in agriculture taken as a whole. Also 'capital goods' imports into the rural areas also began to increase, though at a much slower rate, in the same period.

## DIFFERENTIATITION OF THE PEASANTRY

The differentiation of the peasantry means the break up of the rural populace into a rural bourgeoisie and a rural proletariat. Lenin has said that for Russia, *"two main lines of its develop-ment and outcome are objectively possible: either the old land-lord economy, bound as it is by thousands of threads of serfdom, is retained and turns slowly into purely capitalist, 'junker' economy. The basis of the final transition from labour service to capitalism is the internal metamorphosis of feudalist landlord economy. The entire agrarian system of the state becomes capitalist and for a long time retains feudalist features. Or the old landlord economy is broken up by revolution, which destroys all the relics of serfdom, and large land ownership in the first place. The basis of the final transition from labour service to capitalism is the free development of small peasant farming, which has received a tremendous impetus as a result of the expropriation of the landlord estates in the interests of the peasantry. The entire agrarian system becomes capitalist, for the more completely the vestiges of serfdom are destroyed the more rapidly does the differentiation of the peasantry proceed."* (Preface to the II edition of Development of Capitalism in Russia….. Lenin)

To do so (ie. examine the differentiation of the peasantry and the growth of commodity production resulting from it) we will analyse it under the following sub-heads:

### *Rural-Urban Divide*

Lenin has said that, "the development of commodity economy means the divorcement of an ever-growing part of the population from agriculture, i.e., the growth of the industrial population at the expense of the agricultural population." In explaining the entire process, Marx clarifies, "it is in the nature of capitalist production to continually reduce the agricultural population as compared to the non-agricultural, because in industry the in-crease of constant capital in relation to variable capital goes hand in hand with an absolute increase, though relative de-crease, in variable capital, on the other hand in agriculture the variable capital required for the exploitation of certain plot of land decreases absolutely; it can thus only increase to the extent that new land is taken into cultivation, but this again requires as a pre-requisite a still great growth of the non-agricultural population."

### *Agricultural Labourers vs Kulaks*

Also, the productivity of farm workers in India, on the average, is very low and the instruments of production little developed. And very often, the labourer is not divorced form the instruments of production as he has to use his own implement while working on someone else's farm. One important aspect of capitalism is that wage power must be divorced from the instruments of production.

So though a greater number of the peasantry may be disposed from the means of production and land and though wage labour may have increased with the 'green revolution', the bulk of the labour does not take on the form of regular free wage labour. So the clear differentiation of a rural proletariat and a rural bour-geoisie is yet to coalesce, except in few pockets. The result will be greater amount of labour turning into a commodity coupled withy further impoverishment of the peasantry. Yet this sale of labour power will more take the form of simple commodity production rather than capitalist production. For, say, a middle peas-ant is unable to carry out his farming operations with family labour and so hires three to four workers during the season.

Yet, if the bulk of the produce his family consumes, his exploitation of wage labour gives him no surplus value and so takes the form of simple commodity production even if his produce is a cash crop-say cotton. Much of the labour power sold in the Indian countryside takes this form, and though because of it commodity production expands, the home market grows and there is some change in agrarian relations, but fundamentally it takes the form of simple commodity production with a growth of small-scale farming without generating capitalism or a rural bourgeoisie.

Also, it will be observed that labour-service in the form of share cropping and family cooperation co-exist and continue together with the emerging wage-labour system in Indian agriculture. As regards the extent of growth of the rural bourgeoisie (Kulak) only a general picture can be got from the extent of capital formation in agriculture and the extent of the absorption of the means of production in agriculture. These various estimations are given in other sections. The particular extent of its existence can only be ascertained by specific area studies. The following are some general guidelines to assess whether a farmer comes within the category of rural bourgeoisie or continues as a rich peasant-landlord.

a) The bulk of the surplus must be extracted by the exploitation of labour power. He must employ more wage labour than the labour of his family.

b) The major part of the surplus value generated must be utilised in developing the means of production (machinery) and not in usury, trading, purchase of land, etc.

c) Labour power used must be free wage labour and the labourer should be divorced from the instruments of production-i.e., the implements for working should not be brought by the labourer, but instead provided by the owner of the farm. Payment should be in cash and not in kind.

d) The capitalist farmer must not give his land on rent, on the contrary, he may lease in land.

e) The capitalist farmer must produce for sale. So that bulk of his produce must be sold in the market and not consumed by him and his family.

f) The capitalist farmer should produce for profit-i.e., he must accumulate capital continuously. He must plan his production in such a manner that it enhances profit through agricultural operations.

By taking these six criteria into account we can roughly estimate the class character of the farmer-whether he belongs to the category of rural bourgeoisie or rich peasant-landlord class. Studies may show that a large number of such farmers may be found more to be in a transitional stage not having fully coalesced into a rural bourgeoisie and thereby meeting only some of the above mentioned six criteria.

## *Land holdings-land reforms*

A full ten years after the so-called 'green revolution' the state of operational holdings in the country remained basically the same. So this parcellisation and fragmentation of land is continuing apace, and if, within it, a certain amount of commodity production has been introduced with the help of HYV, cooperative loans, etc., it can lead to only simple commodity production and not capitalist relations of production.

Thus it will be seen that the main trend in agricultural land holding is continuing concentration of land in the hands of a few coupled with extensive small-scale farming. In between lie a class of rich peasantry (15-25 acres) who comprise 3.4% of the households and own 16.2% of land whose number has remained virtu-ally static over the years. It is this section that has, of late, been more vocal as a result of 'green revolution.'

This exposes the glaring existence of labour service in Indian agriculture. The continuing prevalence of tenancy, share-cropping, etc., is a further indication of the perseverance of feudal mode of production. The estimates of cash rent and rent-in-king may not be too accurate and estimates vary from study to study. Other estimates put the rent-in-kind around 25%. But, one important point is that the content of the rent, whether in cash or kind, takes the form of pre-capitalist rent and not capitalist rent, as it involves distribution of surplus product and not of surplus value. The very tenancy acts themselves fix the proportion of distribution of the product between the landlord and serf. If it was to be capitalist rent the distribution would be assessed on the profit generated and not on the basis of the whole product.

## *Household Industry*

Caste-based hereditary labour, according to the 1971 Census, increased over that of 1961 and stood at: barbers 7.11 lakhs, dhobies 9.47 lakhs, fishermen 6.01 lakhs, weavers and spinners 34.03 lakhs, shoemakers 5.97 lakhs, carpenters 12.64 lakhs, blacksmiths 10 lakhs, jewelry and precious metal workers (incl. goldsmiths) 5.76 lakhs, masons 15.35 lakhs, potters 9.66 lakhs, and tanners 0.53 lakhs.

If capitalist industry were to grow and break feudal relations it should first and foremost oust these ancient and backward forms of production and service. But these continue to co-exist, and even expand together with industrial growth. Also a differentiation of the peasantry would simultaneously see the collapse of the household sector, coupled with the growth of modern indigenous industry. Lenin has said that, *"Domestic industries are a necessary adjunct of natural economy, remnants of which are nearly always retained where there is a small peasantry."* (Development of Capitalism in Russia......Lenin)

In India, not only does the caste based household labour continue, but the household industries are propped up by the state and comprador big bourgeoisie.

For example, in the year 1981-82 the total grants plus outstanding loans to the Khadi and Village industries was Rs. 338 crore-Khadi Rs. 208 crore and village industries Rs. 130 crore-given by the government. The very fact that this sector not only continues to exist but is propped up even in the face of expanding industrial production, is a clear example of the compromise struck by imperialism and the comprador big bourgeoisie with the backward pre-capitalist modes of production which they use as a social base for their existence.

### *Home Market*

The differentiation of the peasantry will lead to an expansion of the home market.

In India, as already mentioned, though there has been a continuous increase in food grain production, the percentage finding its way into the market has been static at around 30% to 35%....................A more recent estimate put that no more than the top 10% of the land-owning peasant household has able to produce marketable surpluses on any scale.

Also it will be found that the gross domestic capital formation as a percentage of the total has been virtually static in the decade of 1970s. It increased from 18% of the total (compared to registered manufacturing of 20%) in 1970-71 to 19% of the total (compared to registered manufacturing of 24%) in 1980-81. With little growth in percentage of capital formation in the rural areas it would reflect little growth in capitalist relations there during that period.

Also, during the entire period of the 1950s and 1960s there was a massive shrinking of the percentage of per capita consumer expenditure spent on industrial goods in the rural areas (and also urban areas). In other words there was a relative shrinking of the home market in that period with a larger and larger amount of the families' incomes being devoted to the necessities of life. There was a perceptible reversal of this process in the rural areas from 1968-69 indicating some growth of market from the time of the 'green revolution.'

## INDIAN COUNTRYSIDE

The imposition of Colonial Rule in India led to a drastic break with the past, in that not only did the scale and intensity of the exploitation of the village communities greatly increase, it also led to the introduction of new, and almost entirely parasitic intermediaries between the state and the tax paying masses-intermediaries who were typically induced (or pressurized) to abandon traditional restraint, and discard the old formulae that helped mediate the burden on the typical village peasant or artisan. (It is notable that whereas ancient texts such as the Manusmriti put a cap on agricultural taxes at one-sixth of the produce, and taxes during Mughal rule had risen to something between a third and almost half the produce, during British rule, the peasant was often left with barely a third of the crop.)

Not only were these intermediaries under no pressure to build or maintain traditional water-management structures, the manner in which they were chosen by the British to collect the taxes actually discouraged any diversion of the taxes towards any capital upkeep. Over time, this not only led to the destruction of century-old water management structures, but also to the virtual destruction of the knowledge systems and cultural traditions that had helped build and preserve these water-management techniques over the centuries in states such as Bihar, Bengal, Karnataka, Tamil Nadu and others. (Note that this has been highlighted not only by Rajni-Palme Dutt in *India Today*, but also by several other authors such as Margaret Bourke-White in the *'Famine in South India'* chapter in her book: *"Halfway to Freedom")*. In addition, the British obliged these intermediaries to collect the taxes not in kind, but in cash. Since the peasant had little experience or understanding of a cash economy (or any control on the price of grain and agro-inputs), this put a further burden on the peasant, who now also had to face the thuggery of unscrupulous grain traders and usurious money-lenders-who took full advantage of the highly diminished status of the Indian peasant in the colonial dispensation.

As a result, not only was the typical Indian peasant reduced to a state of utter degradation, social relations within the village also became highly distorted. The traditional solidarity that had existed between villagers was now subject to the divisive and ruinous tactics of the parasitic intermediaries who had the protection and support of the colonial state. As colonial rule progressed, the typical Indian peasant (and village artisan) faced a dual burden.

Not only did the traditional feudal form of exploitation greatly intensify (in comparison to traditional Indian practices), the typical Indian villager was now also subject to the merciless forces of mercantile capitalism, who attempted to extract every ounce of savings or assets that any villager might still possess. Eventually, the typical Indian villager was stripped of all savings, and driven to mortgaging a considerable portion of any assets-whether personal jewelry, land and livestock, or tools and equipment. But even as the Indian money-lenders and grain traders fleeced the village artisans and peasants, their accumulated wealth was not especially significant in comparison to the British mercantile capitalists of the East Indian Company who kept a lion's share of the loot. It is especially important to note that whereas a share of the village surplus went to a few Indian intermediaries (who served as a buffer between the British colonial over-lords and the Indian masses, and generally acted to quash the class struggle), it was the British who expropriated the bulk of the peasant's assets and annual agro-produce, and transferred most of this looted wealth to the British homeland where it was turned into industrial capital and urban development.

This point is of special significance in understanding the unique nature of class contradictions in the Indian countryside (as compared to elsewhere in the non-colonized world.) For instance, almost throughout Europe, the peasantry was also subject to exploitation by the feudal landlords, but the expropriated surplus from the countryside eventually found its way into the urban banking system, and ultimately facilitated in the growth of industrial capitalism. Sections of the impoverished peasantry then had the chance to gradually move into industrial occupations. But India was subject to an especially tragic plight. Its masses were thoroughly impoverished, but their immediate exploiters were able to keep only a very small share of what had been extracted from the peasantry. This meant that even when a transformation into industrial capitalism became possible (after freedom from British colonial rule)-there was simply not enough accumulated savings to fund the rapid expansion of industry that could eventually transform the countryside.

When the British left, they left India with extremely limited material means to industrialize, but a sizeable class of parasitic intermediaries who would serve as a constant drag on future development. Unlike India's pre-colonial mercantile classes and the more nationalist sections of India's traditional feudal nobles and intermediaries (who were mostly neutralized or exterminated in 1857) this new colonial era mercantile class was extremely backward in cultural terms and was generally lacking in nationalist spirit.

In comparison, the mercantile classes in Europe of the late 18th and 19th centuries were not only wealthier, but also more urbane, and displayed a much higher level of cultural, scientific and technological openness and curiosity. This was also generally true of Japan, and to a somewhat lesser degree, of mercantile classes in other parts of Asia that escaped direct European colonial rule (such as in Korea, Thailand and even China). It should also be noted that whereas the mercantile classes in Europe, Japan, Korea and China were moving in a broadly secular direction, the Indian mercantile classes were excessively weighed down by sentiments favouring religious revivalism or extreme sectarianism (a trend that was especially strong amongst Muslims, but also posed a serious problem amongst Sikhs and Hindus).

It is thus important to understand that not only was the Indian countryside weighed down by a new and more intensified feudal oppression in the colonial era (such as through the Zamindari or Mahalwari systems), it was also oppressed by a highly reactionary and socially conservative intermediary class of petty mercantile capitalists who would pose as the biggest obstacle to any radical or revolutionary developments in the Indian countryside.

Finally, it should be noted that colonial rule exacerbated the problems of the Indian countryside in yet another way. Prior to colonization, India was steadily becoming more urbanized, with some estimates suggesting that as much as a third of the Indian population may have been living in large or small towns. In addition, even in the villages, a considerable proportion of the population comprised of skilled artisans-weavers, potters, carpenters, metal-workers, painters etc. The proportion of the population that was exclusively engaged in agricultural production was steadily decreasing. But in the colonial regime, several laws were passed that led to a catastrophic de-urbanization and de-industrialization of India. Trade tariffs and excise duties were set so as to destroy Indian industries, and squeeze domestic trade. In states like Bihar and Bengal, severe restrictions were placed on the use of inland water-ways-causing fishing and inland shipping and transportation to suffer. This led to even greater pressures on agriculture since large categories of highly skilled artisans and non-agricultural workers were thrown out of work.

## EFFECTS OF PARTITION

The problem for India was compounded further by the demands for partition, and the insistence of the Muslim League in splitting the country on communal lines. It is important to note that prior to partition, the best naturally irrigated and most productive lands in Northern India were in North Western Punjab (now Pakistan). Fed by a network of all-weather rivers, no other part of the Indian sub-continent was capable of producing more grain with relatively little capital investment. Similarly, Bangladesh was a much better irrigated land than West Bengal (although it faced far greater risks of flooding than Pakistan Punjab).

But in any case, it is worth pointing out that the bulk of pre-partition India's wheat, cotton and jute were grown in what are now Pakistan and Bangladesh. Not only did partition provide India with a smaller share of the land (relative to the ratios of the Hindu, Sikh and Muslim populations), it also left India without the means to feed itself, and without some of the essential raw materials for industry. (Although India did inherit bulk of the cotton and jute mills, India was unable to take full advantage of this until it had increased its own production of jute and cotton to supply these mills, whereas the raw material producers in West and East Pakistan could at least export their raw materials)

It should also be noted that partition disrupted what had been a somewhat natural distribution of labour between what became Pakistan and India. Since West Punjab and Sindh were better suited to agriculture, the relatively arid states of Rajasthan and Gujarat specialized in all manner of cottage industries and manufacturing that needed the raw materials and markets of West Punjab and Sindh to flourish. In the same way, there was considerable co-dependencies not only between Western and Eastern Bengal, but also between undivided Bengal and its neighboring states such as Assam, Orissa and Bihar. The artificial vivisection of Bengal virtually destroyed the potential of regenerating the traditional trade linkages that had been so vital to the prosperity of the Eastern half of India. Geographically, it should be evident that the management of water and other natural resources cannot be efficiently undertaken without the effective coordination and active involvement of all the states in the Ganga-Brahmaputra region.

But the vitiated atmosphere under which partition took place meant that the possibilities of mutually beneficial cooperation in matters of resource-sharing, trade, transportation and agriculture were practically foreclosed. (To this day, communal forces in Bangladesh have resisted attempts by India to expand cooperation since they would rather see India suffer more rather than see both India and Bangladesh make more rapid progress.) Thus, when the British left, India had become a much more village-based agricultural economy than it had been earlier-it is likely that the British had pushed India back by several centuries in this respect. Yet, its prospects for rapid industrialization were also quite bleak. India had inherited one of the most depressing scenarios in Asia-a fairly densely populated nation with a pitiful urban base, a rural infrastructure in wrecks, a huge mass of population forced to survive exclusively on agricultural production-yet cynically oppressed by petty mercantile capitalists and feudal intermediaries. Moreover, it had also been left artificially divided on communal lines that meant that important historical linkages might never be revived (or revived only with great difficulty).

## SOLUTIONS TO THE PROBLEMS OF INDIAN AGRICULTURE

The subdivision of already small farms in an agricultural model whose profitability is greatly linked to economies of scale makes the survival of the small farmer very difficult. Although the introduction of tractors and other industrial implements, of chemical fertilizers and pesticides, greatly increased the productivity of land and labour, it has not come without considerable burdens for the small farmer who is finding it very hard to compete with large and medium farmers. In most industrial nations, the introduction of such practices in the countryside has been accompanied by a gradual reduction in the rural population so that those who remain on the land can enjoy so

Art, science, and industry of managing the growth of plants and animals for human use. In a broad sense agriculture includes cultivation of the soil, growing and harvesting crops, breeding and raising of livestock, dairying, and forestry (see Animal Husbandry; Crop Farming; Dairy Farming; Forestry; Poultry Farming; Soil Management). Regional and national agriculture are covered in more detail in individual continent and country articles. See also separate articles on the states of Australia and the U.S. and the provinces of Canada.

Modern agriculture depends heavily on engineering and technology and on the biological and physical sciences. Irrigation, drainage, conservation (qq.v.), and sanitation each of which is important in successful farming-are some of the fields requiring the specialized knowledge of agricultural engineers.

Agricultural chemistry deals with other vital farm problems, such as uses of fertilizer (q.v.), insecticide, and fungicide (see Fungicides; Pest Control), soil makeup, analysis of agricultural products, and nutritional needs of farm animals.

Plant breeding (q.v.) and genetics (q.v.) contribute immeasurably to farm productivity. Genetics has also placed livestock breeding on a scientific basis. Hydroponics (q.v.), a method of soiless gardening in which plants are grown in chemical nutrient solutions, may solve additional agricultural problems. The packing, processing, and marketing of agricultural products are closely related activities also influenced by science. Methods of quick-freezing and dehydration have increased the markets for farm products. Mechanization, the outstanding characteristic of late 19th and 20th-century agriculture, has eased much of the backbreaking toil of the farmer. More significantly, mechanization has enormously increased farm efficiency and productivity. See Agricultural Machinery.

Airplanes and helicopters are employed in agriculture for such purposes as seeding, transporting perishable products, and fighting forest fires, and in spraying operations involved in insect and disease control. Radio and television disseminate vital weather reports and other information that is of concern to farmers.

## AGRICULTURE AND FOOD SECURITY

Pre-Independence India suffered repeated famines, drought and food shortages. But following the Green Revolution in the '60s, yields and foodstocks rose manifold. Now, 30 years later, Indian farmers have realised the follies of their tryst with intensive agriculture. Despite 70 per cent of the population being engaged in agriculture and allied activities, declining foodgrain production and access to food remain the two biggest problems confronting the country. Liberalisation has made things worse: commercial crops are eating into the fertile land tracts meant for essential foodgrains. And six years after the World Trade Organisation came into existence, the anticipated gains for India from the trade liberalisation process in agriculture are practically zero.

Agriculture Minister C Subramaniam sat huddled with his senior staff. There was tension in the air. Depending on imported foodgrains coming from 12,000 miles away, particularly when it came in driblets following US President Lyndon Johnson's directive to "teach India a lesson", was humiliating enough. What worried him was that food stocks had reached such a precarious low that they were only sufficient for another two weeks. Still worse, there was nothing in transit. Recalls Subramaniam, "As a last resort, I told my officials and experts to identify the nearest food-carrying ships on the ocean throughout the world. I said we would identify those carrying wheat to other countries and appeal to the US President to divert them to India if other countries could wait for another six to eight weeks."

This was in 1966-67, the critical year of drought when India imported 11 million tonnes of foodgrain. A year earlier, India had imported 10 million tonnes of foodgrain. Around this time, the Paddock brothers, often referred to as 'prophets of doom', concluded in their book, Famine 1975, that by the mid-'70s at least half of India would be led to a slaughter house. Considering that India had a long history of hunger and famine, and the subsequent failure of the government's effort to boost agricultural production in the post-

Independence era, not many disagreed with the conclusive analysis drawn by William and Paul Paddock.

After all, for a country literally surviving on a 'ship-to-mouth' existence, there was little hope. Soon after Independence, India had begun to seek food aid and in fact emerged as the biggest food importer of this century. In 1951, India got ad hoc assistance of 2 million tonnes of foodgrain to tide over the crisis arising from severe drought in several parts of the country. This was the first major food security on a significant scale after Independence.

Before 1947, Indian history was replete with famine, drought and food shortages. Between 1770 and 1880, as many as 27 food scarcities and famines were recorded. At least 20 million lives were lost in India in about 20 famines that had struck since 1850.

Much of this loss was because of misplaced colonial policies which aimed to derive maximum economic gain at the cost of human suffering and misery. After the British had created a transport infrastructure in the first half of the 19th century, they began encouraging farmers to grow crops that could be exported.

The boom in export trade accompanied by rising prices made farmers shift to cash crops like cotton, indigo, poppy and sugarcane. The area under foodgrain subsequently shrank. In other words, efforts to improve agriculture in colonial India were directly linked to the needs of British industries.

In the first half of the 20th century, India's foreign trade showed the same colonial pattern with exports mainly comprising foodgrain and cotton, jute, oilseeds, opium and indigo. Against this, imports consisted mainly of consumer goods. Meanwhile, the cultivator's preference for commercial crops continued.

As is happening now, the trend during the pre-Independence years showed that the best lands available were used for the cultivation of commercial crops. The cultivation of foodgrains was relegated, to some extent, to inferior lands, the yield of which is much less even in normal years.

The result was a steady decline in foodgrain output till the Second World War. Per capita food availability from domestic production too had declined. As B M Bhatia observes in The Famines of India: "India kept the precarious balance between increasing population and declining food production partly by turning from an exporter into a net importer of foodgrains but largely by the easy, though dangerous, expedient of pushing down the already low per capita cereal consumption to still lower levels."

The big gap between minimum requirement and supply continued throughout the post-War period. Till the 1960s, Indian agriculture was not sufficiently able to meet domestic requirements and the country had to rely on foodgrain imports.

## CHARACTERISTICS OF RAINFED AGRICULTURE

In this section we introduce some characteristics of rainfed agriculture that will influence our approach to the problem of deriving recommendations to stimulate rainfed agricultural growth.

As mentioned in the previous section, rainfed and irrigated agriculture coexist in practically every village in India. Public investment programs, however, usually cannot be targeted so precisely.

For practical purposes, they need to be planned and implemented on a larger scale, such as at the village, taluk, district or state level. For example, public programs that provide credit, employ people, or build roads cannot target their efforts to either rainfed of irrigated agriculture; they can only target areas that are relatively more irrigated or more rainfed.

Price policies, on the other hand, can attempt to target rainfed or irrigated agriculture within a given location by targeting crops that may be more likely to be rainfed or irrigated. But few crops are either 100% irrigated or 100% dryland, so some spillover will always remain. Also, every crop is grown over a large geographic area, so it is difficult to isolate the socioeconomic and agroclimatic variables affecting their performance. And while price policies are important, they are not the only approach through which policy makers can influence agricultural development.

## DEFINING RAINFED AGRICULTURE

In the present investigation we use the district as the unit of analysis. We do so for two main reasons. First, through a district level focus our analysis may be relevant for public investment programs that provide infrastructure or other social services to particular areas. Second, a district focus enables us to examine the contributions of both socioeconomic and agroclimatic variables on a nation-wide basis.

The district is the smallest administrative unit for which the required data are available. To arrive at a district-level definition of rainfed agriculture, we consider the percentage of each district that is irrigated or rainfed, and consider predominantly rainfed districts as "rainfed" and predominantly irrigated districts as "irrigated." Obviously there is a certain degree of arbitrariness to any threshold we may choose to distinguish between irrigated and rainfed districts. Several previous studies have faced this same dilemma in categorizing rainfed areas. Some of them have distinguished between irrigated and rainfed districts according to certain criteria such as the amount of rainfall and the level of irrigation.

As mentioned above, all of these definitions suffer from the inability to distinguish between rainfed and irrigated agriculture within districts, but we accept this as an inevitable limitation. Another problem is that both the rainfall and irrigation thresholds are defined somewhat arbitrarily. We discuss rainfall and irrigation thresholds in turn.

## Rainfall Criteria

Bapna et al, (1984) and subsequently Jodha (1985) used broad rainfall thresholds, which is important in order not to be too exclusive. At the same time, the 500-1500 mm range maintains a degree of homogeneity in the types of agriculture under analysis by excluding both very dry, desert areas and very high rainfall areas. Such areas may face unique constraints that limit comparability to agriculture under the more moderate conditions that predominate in most of the country.

Shah and Sah (1993) and Thorat (1993) use narrow rainfall thresholds with a relatively low maximum because they intended to focus on very dry (but not quite desert) areas. It is important to note that the impact of the level of rainfall on crop production is conditioned by both the distribution of rainfall over the course of the season and the factors that determine moisture retention in a given location.

As a result, narrow rainfall thresholds such as those used by Thorat (1993) are likely to combine some areas with disparate moisture regimes and separate others with similar moisture regimes.

A narrow rainfall range probably makes sense only if limited to relatively uniform soil types. Incorporating soil types into the definition, however, introduces yet another variable and makes the definition somewhat clumsy.

For this reason, our preference is to utilize a relatively broad range of rainfall levels such as used by Bapna et al, (1984), excluding only those that are either desert environments or extremely humid.

We choose a range of 450-1600 mm average annual rainfall as the range for predominantly rainfed districts. The lower bound of 450 mm excludes the desert districts of western Rajasthan as well as one district each in Punjab, Haryana and Gujarat. The 1600 mm upper bound excludes the Himalayas, the northeastern states, Kerala, all the coastal districts of Karnataka and Maharashtra, and one coastal district each in Tamil Nadu and Gujarat.

## THE CONTEXT FOR BT ADVENTURE IN GUJARAT

Historical review of the introduction of Monsanto Bt cotton and the emergence of generic products of Bt cotton through Indian companies

1990: Monsanto started negotiating the technology transfer agreement with the Government of India for its Bt cotton package.

1993: The talks failed, as agreement on financial terms of transfer was not reached.

1993: Monsanto started negotiating the technology transfer agreement with the MAHYCO for its Bt cotton package.

1995: DBT (Department of Biotechnology) approved MAHYCO to import 100 grams of cotton seeds containing Bt Cry 1 Ac gene.

1996: Approval by Central Govt. for import of the first Bt cotton variety US Cocker-312. This variety was crossed with the elite Indian varieties to produce locally adapted Bt cotton varieties with Cry1Ac gene.

1996: First trials of Bt cotton.

1997: Feild trials permitted in five states viz., AP, Karnataka, TN, Haryana and Maharastra.

1998: Field trials extended to four more states viz., MP, Rajasthan, Punjab and Gujarat.

1998: Monsanto acquired a 26 per cent stake in MAHYCO, which later became 50-50 joint venture MAHYCO-Monsanto Biotech (MMB).

1998: MMB received approval from Review Committee on Genetic Manipulation (RCGM) of DBT to conduct countrywide-field trials on 85 hectares and to produce seeds on 150 hectares.

1999: RFSTE (Research Foundation for Science, Technology and Education), New Delhi files a Public Interest petition challenging "legality" of RCGM (Review Committee of Genetic Manipulation) under DBT to approve field trials as it is not the concerned authority and also no biosafety regulations were exercised.

2000: DBT allowed Mahyco to conduct extensive field trials, including seed production at 40 sites in six states based on the "total confidential" data from small-scale trials.

2001: MMB approached GEAC (Genetic Engineering Advisory Committee) for commercial release of Bt cotton varieties.

2001: GEAC approved field trials for another year on 100 hectares in seven_states.

2001: GEAC withheld environmental clearance of large-scale cultivation of transgenic Bt cotton in June 2001. Instead it calls for fresh large-scale field trials under the direct supervision of committee set up by ICAR (Indian Council of Agricultural Research) with representatives from MOEF (Ministry of Environment and Forestry), DBT, Dept. of Agriculture and Cooperation and Ministry of Health through Advanced Varietal Trial program of All India Coordinated Cotton Improvement Project. Besides these the GEAC in the press release said "MAHYCO may like to conduct field trials on farmer's field in an area of about 100 hectares under close supervision of GEAC and Monitoring and Evaluation committee. It also advised collection of the complete evidences and data pertaining to impact of transgenic on human and animal food, spread of the cry protein resistant boll worm and impact on non-target soil microflora and other fauna.

October

2001 MMB discovers commercial cultivation of Bt cotton on over 10,000 acres (4000 hac.) in Gujarat, traced the sale of the seeds (Navbharat 151) to Hyderabad based Navbharat Seeds Pvt. Ltd. MMB also discovered that Navbharat has been in the business of selling the seeds for the previous three years and demanded punitive action against the company.

2001 Following this the GEAC ordered the Gujarat Biotechnology Coordination Committee to burn all illegal plantations.

2001 GEAC on insistence from the Gujarat Government ordered that the illegal cotton be procured at a suitable price. But the order was late as the produce was already in the market by that time.

*November*

2001 Case registered with Gujarat High Court against Navbharat for violating the **EPA** (Environmental Protection Act) rules.

*January*

2002 Secretary DBT announces that latest rounds of Bt cotton field trails were satisfactory, based on report from ICAR.

*March*

2002 Centre approves three years commercialization of three Bt cotton varieties – Bt MECH-12, Bt MECH-162 Bt and Bt MECH-184 with a set of conditions. It disapproved MECH 915 Bt cotton seeds.

*May*

2002 Rasi seeds Company Ltd. gets nod from DBT to conduct trials.

*August*
2002 Karnataka government bans sale of Bt cotton seed temporarily.

## A BRIEF LITERATURE REVIEW OF THE CURRENT BT COTTON DEBATE

Technological change in tropical developing countries has been studied from institutional, socio-economic and cultural, policy and structural perspectives. Incentives for farmers, it has been realized, are the prime movers for technological change in most cases. These incentives or benefits may sometime be high in short term but may have low returns in the long term (as in the case of chemical intensive approach to agriculture).

From the portfolio perspective, risk and return are two obvious determinants of this change. There can be four combinations, high risk, high return; high risk, low return; low risk, high return; and low risk, low return. In the case of Bt, the risks associated with the technology are

(a) *environmental:* When the pest infestation goes down, the possibility of birds and other predators getting attracted goes down. To that extent, there is an impact on the avian biodiversity. Whether this effect is lesser or more than the pesticide sprayed cotton is a matter for further investigation. The possible effect on soil microorganisms, on other species through diffusion of Bt gene including weeds are other environmental implications.

(b) technological, i.e., resistance development of pests to the Bt gene. Such risks associated with resistance are being investigated. There does not seem to be conclusive evidence. Some argue for precaution on this ground.

(c) consumer health such as possible allergenic role of the food products (in this case, the oil of cottonseed). These risks have not been investigated adequately in the Indian context

(d) *socio-economic:* The socio-economic effect on labour by way of suspected decrease in demand, has recently been found to be offset by increased duration of the crop from six to nine months. This increased demand for labour used in picking cotton. The labour finds their productivity going up in Bt cotton compared to conventional hybrid cotton. Reportedly the number of balls infested with pests being higher in the later case makes the task of picking good quality cotton bit tedious. The labour contractors also reportedly benefit because of the significantly higher volume of cotton being picked up by the labour everyday enhancing their own margins.

(e) ethical, implying doing things with inadequate information about right or wrong outcomes on various stakeholders, imposing in the process, costs on future generations. The ethical issues are extremely serious not only because of the decisions taken with inadequate information but also because of the possible hazards in natural ecosystem. The constrained choice of future generations in terms of cotton varieties without Bt gene also involve a moral dilemma. Robinson recalls that every technological revolution in the past has had major impact on the perceptions of the public about right and wrong and consequently the legitimacy of technological change. The discourse on usefulness of genetic engineering will have to be guided by better science but also more widespread debate. Scientists cannot complain that their confidence in the technology is being questioned by people whose understanding of science is limited. In fact, the farmers who have crossed Bt cotton variety released illegally with the local hybrids and have found economically beneficial results are paradoxically not worried about environmental implications. Most of the farmers we talked to seem to have faith in the ability of the scientists not only to monitor that but also to educate them as and when situation so warrants. The fact that scientists have delivered better hybrids in the past to overcome the resistance as well as other problems seems to assure the farmers that they would get the answers to any possible problem that transgenic cotton might create in future. The irony is that scientists themselves are not doing much systematic monitoring of the large-scale farmer participatory Bt research going on in Gujarat. For once, the scrutiny of science is not being matched with the popular faith in its capacity to deliver. The regulatory framework regarding biotechnology in India is summarized in Annexure one. It is a different matter that popular faith in this framework is not as strong as one would expect in a democratic society.

Regulatory efforts with regard to Bt cotton in India and quoted officials of the Department of Biotechnology admitting their lack of knowledge regarding the long-term safety. Scoones believes that by not engaging the critics in the science policy network, 'a sense of disengagement, distance and distrust within the regulatory process emerges.' Many agricultural and ecological scientists have felt excluded from the regulatory process. He quotes senior scientists involved with DBT about the doubt they have in the ability of DBT to enforce their own guidelines. He argues for some forms of accountability outside the science – industry – policy networks to make current discussion on biosafety regulations meaningful.

Herring, though criticizing the nature of dialogue between policy makers and the disadvantaged communities nevertheless wonders whether scarce resources are to be spent in building a more strict, widespread and bureaucratic regulatory regime or if these resources ought to go to public health, education or extension. Excessive caution, he suggests, may drain the resources required from alternative uses. He adds that insufficient caution might generate unanticipated harms.

Kinderlerer and Adcock raise the question of justice involving introducing new technologies and imply that if poor benefit, then risk might be worth taking.

In Gujarat, small and large farmers (with or without effective irrigation) seem to have voted for Bt, including the illegally marketed one, disregarding other consequences. Does it make it just? Do ends justify the means?

Gupta and Sinha have asked this question in the context of environmental governance and have wondered whether the ethics of not investing in alternative ways of solving a problem makes even more profitable outcomes from a particular approach, more legitimate and moral. Thus, if we ignore IPM and we assume that Bt cotton gives more profit than IPM approach, will Bt approach become more ethical just on that ground, regardless of overall consequences for environment.

Bharathan believes that in the case of Bt cotton, the democratization of knowledge, technology and discourse would require greater participation of the scientists from different disciplines. Venkateshwarlu had argued that despite general governmental sympathy for the farmers who violated GMO regulations, the real solutions lay in creating wider awareness about biosafety regulation rather than only making regulations more strict.

In a recent report, Jishnu referred to seed industry estimates that nearly half of the Gujarat's fields were planted with Bt cotton. She also looked at government figures about seed supplyfrom public sources, and from legal channels and the acreage under cotton to estimate the area under illegal Bt cotton.

Farmers in Gujarat have no qualms to say that the issue is not whether Navbharat Bt or not, the issue is how to get all the seed that one needs one way or other. Seldom has a technology found so widespread approval from the users without any intervention of formal extension machinery.

Undoubtedly the politicians have supported it by expressing helplessness and claiming inability to get illegal cotton uprooted by paying compensation to the farmers. While Jishnu refers to the advantage of Navbharat Bt cotton in terms of shorter duration, farmers have gone a step ahead and found that by extending the duration for about three months, they can almost double the yield with marginally extra cost. Navbharat Seed Company sources have maintained in personal communication (March 10, 2004) that they have not sold any seed for last two years.

By implication, the entire diffusion of Navbharat 151 Bt cotton seed in Gujarat during the last two years has been achieved by farmers, traders and other seed companies, which have produced seed one way or the other incorporating the Bt gene from Navbharat 151 variety.

What is even more interesting is that several postgraduate students of Gujarat Agricultural University have become entrepreneurs by developing the seed at their private farms and thus earning while learning.

Whether their values will get so shaped in this process of illegal seed production that after twenty years when some of them would be head of the department or members of regulatory bodies, they would have perhaps less difficulty in living with legal or ethical violations. They might not even ask question about ethics in such situations.

The discourse on Bt cotton is being outpaced by the fast changing ground realities. Sahai referred to the paper by Quaim and Zilberman , which created lot of controversy. It stated that the genetically modified cotton having Bt gene had shown higher yields in the experimental plots in India. Sahai argued that the failure of Bt cotton in Maharashtra and Andhra Pradesh contradicted, "the exuberant projections of two foreign scientists publishing from an American university...".

She argued that the motivations of the editorial committee of *Science,* a very reputed journal could be considered suspect. We are worried that if such be the logic, would widespread satisfaction with Bt cotton in Gujarat during last two years vindicate the decision of editorial board of *Science.*

It is a different matter that farmers have found Mahyco Mansanto Biotech (MMB) cotton yielding higher than Navbharat 151 seed but their preference still remains for the latter because of its ability to yield for longer duration, earliness in flowering and almost one fourth the seed cost (MMB seed costs about 35 USD per 450 grams whereas Navbharat costs nine to twelve dollars for the same quantity)

# 4

# Biotechnology in Agriculture: Processing, Product Quality and Uses

## INTRODUCTION

In this chapter, biotechnology is defined as "any technique that uses living organisms to make or modify a product, to improve plants or animals, or to develop microorganisms for specific uses" (after the Office of Technology Assessment, US Congress). The document focuses on the development and application of modern biotechnology which is based on the use of new enabling techniques of recombinant-DNA technology, often referred to as "genetic engineering".

Biotechnology has great potential to influence and benefit agriculture, forestry and fisheries. Modern techniques of biotechnology offer the potential of moving any cloned gene from any organism into any other organism and confer much greater precision and speed in achieving results as compared to conventional techniques. In conjunction with conventional technologies, modern biotechnology holds promise of increased and sustained productivity, efficient processing for improved product diversification and utilization, adaptation of product quality to functional requirements, and decreased reliance on agrochemicals and other external inputs. It may also promote better conservation and use of genetic resources, and environmentally friendly management of natural resources. However, the number of marketable products and their influence at the farm level still seems to be limited, but is likely to increase in the next decade.

Biotechnology also poses certain challenges. These are largely determined by how, where and when it finds application. In general, the fast-paced research, predominantly funded by private-sector investment, and use of intellectual property rights in industrialized countries are seen as evidence that the application of biotechnology will hold the key to competitiveness and comparative advantage in many fields, including agriculture and food. Biotechnology, with its vast potential and challenges, is thus of the utmost importance to agricultural development, and hence to FAO. The science of developing biotechnological tools cannot make a distinction between developing and developed countries. However, the application of such tools in the development process requires preconditions which are easily found in many developed countries, but which hardly exist in most developing countries. The current paths of research and development have given rise to concern that the disparity in harnessing biotechnologies for agricultural and economic development may increase between industrialized and developing countries.

The preamble to the FAO Constitution requires that the Organization work towards raising levels of nutrition and the standards of living of all peoples, securing improvements in the efficiency of the production and distribution of all food and agricultural products, improving the conditions of rural populations and thus contributing towards an expanding world economy. Therefore, one of the major tasks for the Organization is to ensure that the benefits of biotechnology will be shared by people in the north and the south, in both large and small, and rich and poor countries.

## DEVELOPMENTS IN BIOTECHNOLOGY

There exists a gradient of biotechnologies, depending on the degree of sophistication, complexity, stage of development and application. The lower side of the gradient comprises simpler but widely used techniques such as in vitro culture, rhizobium technology, fermentation. The upper side of the gradient includes advanced techniques involving genetic engineering. The gradient of biotechnologies may be matched with the gradient of national capabilities, economic investment and efforts to provide the possibility of choosing the appropriate techniques and approaches with the most positive impacts. The pattern of use and development of biotechnologies in developed and developing countries thus varies considerably.

## STRATEGY OF BIOTECHNOLOGY

A balance between biotechnology and conventional research is crucial to CGIAR. The strategy of CGIAR crop improvement centres is to use relevant, tested biotechnology techniques for more efficient execution of their research agendas.

The centres are interested in developing new tools for their own work as well as for their national programme partners. In some cases, new diagnostic, analytical, or plant breeding tools are developed by the centres to improve their own research capability in meeting critical international needs. In other cases, new tools are made and adapted specifically for national programme research efforts. A well-established linkage strategy of IARCs, that of using cooperation to pass on useful research techniques to national programmes, will also be a way forward in biotechnology, particularly for smaller NARS. One route may be to link IARCs with developing-country universities as future centres of strategic research. As skills mature, the universities can then exploit IARC connections to begin their own interactions with advanced institutions and companies in the industrial economies.

The "in-house" management of biotechnology in IARCs is rooted in this strategy and is still in a formative stage. Centres each have one, two or sometimes three scientists who act as coordinators of biotechnology research efforts. They listen to programme needs in demand and supply consideration together with the help of senior managers who deal with increasing concerns regarding contractual, legal and safety matters. In some centres, individuals have been the catalysts in making biotechnology research operational. Several of the larger centres have established separate biotechnology units.

## BIOTECHNOLOGY RESEARCH-PAST AND PRESENT

Biotechnology is driving a global transformation from the treatment of illness to the treatment of wellness; at the same time, the changing world is transforming biotechnology. The biotech industry is no longer centered in the United States and Europe: its maturity means that competition for resources (both human and technological) is increasing as the number of countries supporting viable life sciences industries grows. Nearly every part of the world is looking toward a future dramatically affected by biotechnology. The term "global transformation" refers not only to geography. The transformation is pan-industry, and it extends across technology and the innovations that result.

Biotechnology research in CGIAR and its potential applications were first discussed by the Group in 1981 (CGIAR, 1981). At that time, with the exception of the research of the International Laboratory for Research on Animal Diseases (ILRAD) which was well under way, mostly tissue culture techniques were being used by CGIAR scientists for multiplication and production of disease-free plant material, and genetic engineering techniques were discussed in a futuristic sense. Still, it was predicted that genetic engineering would not replace traditional plant breeding techniques, but would provide new tools for breeders. As science progressed, an overview was presented to a CGIAR mid-term meeting on the activities of biotechnology research in CGIAR in 1985; this overview was updated in 1988 (Plucknett, 1988). By then it was clear that "biotechnology brings new tools, new ideas and new approaches to agricultural research. In most cases, biotechnology will not supplant what, for want of a better term, might be called conventional agricultural research.

Rather, biotechnology will be used in conjunction with traditional plant breeding and other agricultural research to help provide new information and tools to solve problems." The present chapter draws extensively on CGIAR thoughts on developments in biotechnology since 1989 in a paper prepared for the World Bank/ISNAR/Australian study, case-studies of IARC activities, and recent papers put forward by the Technical Advisory Committee (TAC), the Centre Directors and the CGIAR Secretariat on the emerging socio-economic aspects and future issues raised (such as biosafety and intellectual property rights). Biotechnology includes both scientific advances and technologies that can be developed from those advances. For an advance to become useful, it must be packaged in a usable way. For example, hybrid maize is a technological delivery package for scientific advances that emanated from a new understanding of basic genetic principles, leading to the development of hybrid seeds that farmers could plant, to their benefit. An understanding of the relationship of science and technology can help us to see the potential benefits of biotechnology in the international agricultural research centres and their national programme partners.

CGIAR believes it can help in two fundamental ways: by providing a bridge for the flow of needed information and germplasm (seed and plant material) between developed and developing countries, and by ensuring that the agricultural needs of the developing countries are not lost in the total research effort, given that there would be little money to be made by private firms in the improvement of orphan commodities or in many technologies designed for resource-poor farmers. Most IARCs that deal with crop or livestock improvement are involved, or are likely to become involved, in biotechnology research. This includes cellular and molecular biology and the new techniques derived from these to improve the genetic makeup and management of crops and animals.

However, few IARCs will develop new basic knowledge in support of biotechnology, but rather will become involved in finding applications for new scientific advances. Thus IARCs are dependent upon scientific advances and resulting technologies in efforts to adapt technologies to developing country problems. In a sense, IARCs are toolmakers in biotechnology, using basic scientific knowledge to develop useful technologies in crop and animal research (Plucknett, 1988). If IARCs are to play an important role in adapting biotechnology tools to the needs of developing country agriculture, new methods for acquiring technologies must be considered. It is true that commerce drives many of the applications of genetic engineering (Scowcraft, 1989).

Not only must the centres themselves be able to negotiate creatively with the private sector, but also appropriate research institutions in the developing countries may need to do so. Several developing countries' research institutions have already recognized the need to apply state-of-the-art technologies to their ongoing research efforts. At present, the scale of CGIAR biotechnology research is modest-especially compared with that of the private sector-and probably will remain so. In 1990, for example, CGIAR plant and animal biotechnology research totalized US$14.5 million.

This was less than 5 percent of the total CGIAR budget and only a traction of the US$55 million spent on agricultural biotechnology by one US corporation, the Monsanto Chemical Company. Half of the CGIAR outlay on biotechnology went to plant research, distributed across nine international centres working on 15 crops.

The other half went to animal research, primarily at ILRAD in Nairobi. This centre conducts state-of-the-art molecular biology research for the control of two major livestock diseases: tick-transmitted East Coast Fever (theileriosis) and tsetse fly-transmitted African sleeping sickness (trypanosomiasis).

## THE ROLE OF IARCS IN ADVANCED RESEARCH NETWORKS

Many international centres play a crucial role in advanced networks that link centres with advanced basic science institutions. Such networks help to target research, develop methodologies, establish new research channels, expand awareness of centre-based research, and provide a means to acquire new technologies. The centres gain because they can link their own programmes with those of advanced and developing country institutions on problems of mutual interest. Collaborating advanced institutions benefit because they can gain an understanding of major developing country problems that might benefit from biotechnology approaches, as well as learn more about the genetic resource collections held by the centres. The oldest such network, as far as the centres are concerned, is the Rockefeller Foundation-supported network on biotechnology in rice, established in 1985. This network links advanced laboratories in Europe, the United States and elsewhere, with the international Rice Research Institute (IRRI) and the International Centre for Tropical Agriculture (CIAT). Molecular maps are being developed to allow breeders to determine whether an individual rice plant contains known genes. Many potentially useful genes have been cloned from rice, including the gene "oryzacystatin", an inhibitor of the digestive enzymes of insect pests. More will be said about this important undertaking later in this chapter.

The formation of the Cassava Biotechnology Network in 1988 at CIAT, involving scientists in Latin America, Africa, Europe and the United States, brought new techniques to difficult research problems in cassava. A major effort involves studying the biochemistry and genetics of cyanogenesis (the formation of HCN, hydrocyanic acid, in plants). Social scientists will assist with farming systems studies in areas where cyanogenic cassava is grown. Findings will help define research objectives: whether to eliminate cyanide throughout the plant, or to increase specific enzyme activity to reduce the presence of cyanogenic compounds. A further goal is the production of true seed as a joint venture between the International Institute of Tropical Agriculture (IITA) and CIAT.

The International Potato Centre (CIP) and the International Centre for Maize and Wheat Improvement (CIMMYT) are also effectively using networks. CIP, which works with potato and sweet potato, has established an extensive network of collaborating institutions in the United States, China, Israel and Europe for activities in genetic engineering and RFLP analysis. Efforts such as this have enabled CIP to obtain access to scientific expertise in a cost-effective manner. CIMMYT was involved in establishing and participating in a maize RFLP network under the EUREKA funding and guidelines of Europe. This network consists of public and private institutions in France, Germany, the Netherlands and Italy. Their goals are to utilize molecular markers (RFLPs) to complement the genetic map of maize; to fingerprint European and CIMMYT maize inbred germplasm; to identify (map) Quantitative Trait Loci (QTLs) for important agronomic characters; and to develop strategies for using marker-assisted selection of breeding programme. CIMMYT also participates in the International Triticeae Mapping Initiative (ITMI), a network of scientists interested in mapping the wheat and related genomes.

In addition to the research networks, ISNAR established an Intermediate Biotechnology Service (IBS) in early 1993, to serve as an advisory network on management and policy aspects of biotechnology research in developing countries.

CIAT-Biotechnology research began in the period from 1980 to 1984, with emphasis on tissue culture for conserving cassava germplasm and accelerating rice germplasm improvement. In 1985 the Biotechnology Research Unit (BRU) was established. BRU focused further on critical challenges in CIAT's crop germplasm and expanded to include the study of selected microorganisms and modern biochemical and molecular techniques. Research is carried out through special project funding and includes:

- Pilot projects with IPGRI (the former IBPGR) on an in vitro genebank and cyropreservation. This work was completed in 1991.
- Gene tagging and mapping of Rice "Hoja Blanca" Virus (RHBV) and rice blast resistant genes (funded by the Rockefeller Foundation, this project is ongoing).
- Construction of a cassava molecular map, to be completed in 1994 (RF funding).
- Construction of a molecular map of tepary beans to be finished in 1994, funded by Belgian AGDC.
- Operations and coordination of the Cassava Biotechnology Network (CBN), 1992-1997, funded by Dutch DGIS.

CIAT also established several collaborative research projects with developed country institutions in the United States and Europe to work on, inter alias, genomic studies in cassava, rice biotechnology and molecular markers for evolutionary studies in common bean.

Additional research includes the characterization of mechanisms involved in resistance and tolerance to biotic and abiotic stress in plants. This work can be categorized as: (1) the characterization of resistance mechanisms to pests and pathogens (e.g. resistance to bruchids, a major pest in beans, and developing the molecular bases for co-evolution in bean pests); (2) characterization of physiological and biochemical processes in plants and bacteria (e.g. drought and heat tolerance in cassava, aluminium tolerance to acid soils, and cassava starch fermentation); and (3) development of methodologies. As technologies are developed, CIAT shifts activities from BRU into the appropriate research programme, which then continues the work. Since 1988 six major projects have been handled in this manner (CIAT, 1992).

In addition to research efforts, CIAT has trained nearly 90 people from 20 countries in Latin America, Asia and Africa in tissue culture and the biochemical and molecular characterization of genetic resources. Future training will emphasize graduate studies and in-service research, linked as much as possible to national research programmes.

CIMMYT-Since the early 1980s, CIMMYT has utilized various tools of biotechnology in wheat and maize improvement. They have included the use of tissue culture for producing the inter genetic hybrid triticale, somaclonal selection for salt tolerance in wheat, and the enhancement of recombination in the wide-hybridization of wheat with wild relatives, and maize with Tripsacum. Enzyme-Linked Innunoserbent Assay (ELISA) assays were and are being used in the routine diagnosis of several viral diseases, including barley yellow dwarf virus (BYDV). Various protein and isozyme markers have been used in detecting chromosomes of wheat in evaluating protein quality of wheat varieties.

With the completion of its new applied biotechnology facilities in 1989, CIMMYT's efforts in biotechnology were enhanced to include the use of molecular marker technology for location and manipulating the genes involved in traits of agronomic importance. These efforts are the focus of the Applied Molecular Genetics Laboratory. More recently, the Tissue Culture Laboratory was established in order to take advantage of recent reports of successful maize and wheat transformation. All activities in biotechnology seek to enhance the breeding programme, through either more efficient selection techniques or enhanced germplasm. Technology evaluation and adaptation for use at CIMMYT and in its client countries, training and technology transfer and collaborative research, all play a part in these efforts. The Applied Molecular Genetics Laboratory has concentrated in the last few years on establishing efficient protocols for the use of marker technologies in wheat and maize. There are four aspects of protocol development:

- use of non-radioactive DNA hybridization techniques to provide a safer working environment;
- adaptation of RFLP and other marker techniques for large-scale analysis;
- optimization of cost-effective experimental strategies; and
- development of software for fingerprinting databases, molecular data capture and verification and cost analysis.

Current collaborative research in maize includes the evaluation of the genetic diversity of tropical, subtropical, highland and African germplasm. The current database contains molecular fingerprints for over 100 lines. This information has enabled CIMMYT's breeders to understand better the heterotic groupings and to develop more appropriate testers for germplasm improvement. A major study involves the genetic base of resistance to multiple corn borers using RFLP genotyping and field evaluations of populations segregating for resistance. Results indicate resistance is polygenic and gene action is primarily additive. It is expected that RFLP markers will facilitate the incorporation of resistance into elite germplasm, increasing breeding efficiency and reducing costs.

The application of marker technologies in wheat is more complex, because of its larger genome and its polyploid nature. There has been progress in pilot mapping studies using large-scale RFLP analysis for durable leaf rust resistance and resistance to bacterial leaf blight. Projects are being developed to widen the genetic base of modern wheat through marker-assisted introgression of desired traits from wild bread wheat relatives (e.g. resistance to viruses, fungi and bacteria; and tolerance to abiotic stresses such as salinity, drought and aluminium). Molecular markers are expected to become routine breeding tools at CIMMYT; this will require a "service-oriented" laboratory to handle the expected large numbers of routine molecular analyses.

The Tissue Culture Laboratory seeks to enhance the resistance of tropical maize to major insect pests through transformation which offers breeders a new tool for broadening the genetic composition of germplasm. Several major tropical and subtropical CIMMYT lines are being evaluated for embryogenic callus formation and regeneration potential. Also under way are insect bioassays of partially purified toxins from Bacillus thuriengensis (Bt) strains collected in Latin America in collaboration with the Mexican plant molecular biology institute, CINVESTAV, and the EMBRAPA maize breeding station in Sete Lagoas, Brazil. Efforts in maize transformation as well as collaboration in several developing and developed country institutions will soon be enhanced, with the recent approval of a special project under UNDP funding.

CIP-Recent work expands on the new Restriction Length Polymorphism (RFLP) map for potato produced by Cornell University, in collaboration with CIP. A total of 220 markers are now available. Randomly Amplified Polymorphic DNA (RAPD) markers, generated by the Polymerase Chain Reaction (PCR), have been adapted to analyze DNA variation in tuberbearing Solanum, demonstrating the RAPD markers are suitable for large-scale assessment of the genetic diversity of potato populations.

Transformation through the insertion of antibacterial protein genes and the coat-protein gene for Potato Leaf Roll Virus (PLRV) into potato clones is now possible. Two genotypes, Desiree and 86007, were used for transformation with Agrobacterium tumefaciens and A. rhizogenes as vectors for gene coding for antibacterial proteins.

CIP's virus detection work continues to provide antisera to the major potato viruses, and a new PLRV antiserum is being prepared. Virus detection kits for potato viruses are a widely used CIP product. More than 750 000 virus samples were evaluated through the distribution of antisera, Das-ELISA, and NCM-ELISA kits. There is a great demand for this technology in all countries with severe Sweet Potato Weevil (SPW) problems since progress has been slow in developing SPW-resistant cultivars. The recent discovery of Bt strains effective in controlling SPW has opened new avenues for genetically engineered sweet potatoes with the Bt-toxin gene. In the near future transgenic sweet potatoes with resistance to SPW may become available.

To develop transgenic potatoes with resistance to tuber moth, CIP is collaborating with the Comitato nazionale per la ricerca e per lo sviluppo dell'energia nucleare e delle energie alternative (ENEA) in Italy, Michigan State University (MSU), and Plant Genetic Systems (PGS) of Belgium. ENEA has focused on the development of two DNA sequences similar to other known Bt sequences. These were used for transforming potato plants by Agrobacterium infection. Collaborative work between MSU and PGS has produced several potato cultivars with resistance to PTM. Several CIP advanced clones have now been selected for engineering Bt genes with resistance to PTM. Other genes of interest for transformation efforts include the Cowpea Trypsin Inhibitor Gene (CPTI).

ICRISAT-ICRISAT scientists use biotechnology tools to overcome crop production constraints, but only where conventional techniques cannot be applied, or where biotechnology is more efficient or cost-effective. ICRISAT adapts proven biotechnology techniques to its research. In cereal biotechnology research, ICRISAT aims to use molecular markers in crop improvement. In collaboration with the University of Milan, a RFLP map of sorghum is being developed using maize probes. Similarly, with the collaboration of scientists from the United Kingdom, a RFLP map of pearl millet is nearing completion. Refined molecular maps for both crops will enable researchers to make rapid progress in their improvement. ICRISAT participates actively in global efforts to use molecular markers in crop improvement.

Some transformation work is also being pursued. Whole plants were regenerated from various organs or single cells of pearl millet, and from shoot apices of sorghum. Future work will build upon collaboration with public institutions in the United States and elsewhere, particularly in sorghum. In collaboration with the Scottish Crop Research Institute (SCRI), United Kingdom, a molecular map of groundnut is being developed to identify markers for agronomic traits. This will help identify the coat protein gene of the Indian peanut clump virus and make plasmid constructs suitable for the transformation of groundnut.

Research in chickpea and pigeonpea has emphasized reliable tissue culture and plant regeneration techniques for use in wide crossing and transformation. Interspecific hybridization of chickpea (Cicer arietinum) and a wild Cicer spp. is in progress. Target genes are those conferring resistance to insects and to fungal pathogens. For pigeonpea, wild species hold many desirable characters not present in the crop. An interspecific hybridization programme involves crossing Cajanus cajan with Cajanus platycarpus, and overcoming current barriers to hybridization.

IRRI-As mentioned earlier, IRRI's biotechnology work has been supported by the Rockefeller Foundation International Programme in Rice Biotechnology, established in 1985.

This network serves as a prototype for successful interactions between developed and developing country institutions, and includes the participation of IRRI, CIAT, IFPRI and ISNAR.

Of the total amount invested or committed to the programme by the Rockefeller Foundation since 1985 over US$48 million-slightly over US$22.5 million have gone to institutions in North America, Europe and Japan; nearly US$19 million to developing countries-most of the funds being spent on 130 fellowships; US$ 1 million towards information dissemination; and slightly over US$6 million to international agricultural research centres.

Of the funds going to IARCs, US$5 million supported 12 biological research projects, and US$ 1 million supported six social science research projects (Rockefeller Foundation, 1992).

The rice biotechnology programme aims to strengthen the capacity of research institutions in developing countries to assume increased responsibility for further development and application of rice biotechnology. IRRI, in collaboration with developed country laboratories, has worked to bring about some of the major highlights of the programme, including the following (Toenniessen, 1992):

- a molecular genetic map of rice with over 600 markers has been developed and distributed around the world and is now being used to tag important genes;
- rice transformation has been achieved for major tropical (indica) rice-breeding lines such as IR72. Potentially useful transgenes, such as those for virus and insect resistance, have been transferred to rice and are currently being tested;
- DNA markers of rice pests and pathogens have been used to determine their genetic diversity and structure; markers have helped breeders to identify resistant germplasm and to breed for durable resistance;
- developing country scientists have participated in all research projects.

Information and technologies generated are passed to developing country institutions for further research and implementation.

ILRAD-ILRAD's work represents approximately half of CGIAR funds invested in biotechnology research. Recent accomplishments include joint work with the University of California at Berkeley which led to a completed map of the entire genetic material of Theileria parva, an important parasite of domestic African livestock, which causes the disease known as East Coast Fever. This restriction map is the first complete map made of the genome of a protozoan parasite. Techniques involving modern molecular biology, biotechnology and genetic manipulation were used in the mapping exercise. The parasite genome map will be used to monitor the occurrence of genetic recombination in the parasite's life cycle, which involves the generation of a wide variety of parasite strains. Knowledge of the basis of this diversity will help researchers develop a vaccine that will work broadly and effectively against all strains.

The theileriosis programme continues to develop and refine antibody-based ELISA systems and DNA probes for precise identification of tickborne disease organisms. In this effort ILRAD works closely with laboratories in Australia, United States and Latin America (ILRAD, 1992). ILRAD has identified, characterized, cloned and expressed a surface antigen of the sporozoite stage of T. parva. The recombinant protein when combined with adjuvant is capable of inducing protective ability has also been identified, characterized, cloned and expressed, and its immunizing potential in cattle is now being tested. These antigens represent the basis for the development of new, genetically engineered vaccines against East Coast Fever.

The trypanosomiasis research programme has developed highly sensitive tests (monoclonal-antibody-based ELISAs) for distinguishing trypanosome species. The reliability of this work was validated in 1991 by researchers in eight laboratories in Africa whose work was supported by FAO and the International Atomic Energy Agency (IAEA) in Vienna. The utility of these tests in diagnosing human trypanosomiasis (sleeping sickness) is now being tested with the World Health Organization in Geneva.

ILRAD and its partners in a network of laboratories in the United Kingdom, Switzerland, Israel, the United States and Australia, are engaged in developing a physical map of the bovine genome. In this collaboration, ILRAD provides a major genetic resource by breeding F1 and F2 families derived from crosses of trypanosusceptible and trypanotolerant parents and determining their trypanotolerant status. ILRAD has also developed technologies and primers to define cattle populations and subpopulations, known as RAPDs, which will be highly valuable in the description and conservation of global bovine genetic resources.

## TRAINING AND INSTITUTION BUILDING

An important way to diffuse research technology is through training and institution building. CGIAR has an illustrious record in research training to help strengthen developing country capabilities. Between 1985 and 1989, CGIAR-supported centres conducted training courses, most of them lasting from two to four weeks, to an estimated 25,000 developing country researchers and scientists in all areas of crop and livestock improvement (CGIAR, 1991b). In addition, research fellowships in biotechnology have been offered at several centres.

Most IARCs are hesitant to promote biotechnology training in countries without significant plant breeding programme and without appropriate laboratory facilities (Collinson and Weight Platais, 1992). Experience shows that researchers returning to a national system, after receiving training from a developed country institution but with little or no means of mobilizing their skills, become frustrated and often feel "cut off" from scientific advances.

On the institutional side the centres are increasingly being asked by national agencies to help set up creative programmes that often have a biotechnology component. In 1986, for example, the Andean Development Bank (CAF-Corporación Andina de Fomento) asked IARCs in the Latin American region to assist in some of the activities of its recently created biotechnology programme. In response, CIP began work with CAF on projects in five Andean countries (Peru, Bolivia, Ecuador, Colombia and Venezuela), which range from using tissue culture for disease-free material, to in vitro potato seed production, to facilitating the transport of planting materials to remote areas.

## INFORMATION DISSEMINATION

As research advances are made in biotechnology, the importance of information flow becomes essential. IARCs help NARS by providing technical information which may be otherwise difficult to obtain. Information resources currently available include Ag-biotech news and information, published by CAB International. The CGIAR Biotechnology Task Force (Biotask) was formed in 1988 to help raise the awareness of biotechnology in all elements of the CGIAR system, including IARCs, donors, TAC and national partners. In addition to providing a forum for discussion on issues of biotechnology, other activities were undertaken, such as assistance by the Dutch Government in distributing the ag-biotech journal to 120 developing country libraries. Biotask ended its work in 1992.

In regard to information needs, in 1990 the Directors-General of the CGIAR centres stated: "The fullest application of biotechnology may be constrained by non-technical issues such as biosafety regulations, intellectual property management and public awareness. The dissemination of balanced, accurate information concerning biotechnological activities carried out in the Centres and elsewhere in the scientific community in the service of agriculture would ease such constraints" (CGIAR, 1990).

## CURRENT AND FUTURE POLICY CONSIDERATIONS

### Biosafety

From the earliest stages of research on specific GM organisms, scientists and policy makers have attempted to design and implement regulatory processes that ensure their proper safety assessment. Biosafety processes formalized in the Cartagena Protocol on Biosafety have now become a pre-requisite for GM research and any release of GM products into the environment. The Biosafety Protocol focuses on the potential effects of GMOs on the environment, although most biosafety regulatory systems have broadened to include food and feed safety and, in some cases, socio-economic considerations and ethics. Most biosafety systems are sequential learning processes in which the decision whether to advance a product to the next regulatory stage uses knowledge and data accumulated up to the point where the developer submits the application dossier. The developer can generate data and knowledge in country, use data generated elsewhere, or in some cases use data from a related crop (i.e., potato and sweet potato). Countries differ in the extent to which they accept data from other countries. The task of the decision maker—such as a national biosafety or other competent authority—is to decide whether the data submitted are sufficient to demonstrate an established level of safety.

Biotechnology regulations in the developing world are complicated, partly because of a diversity of players in biotechnology research and testing, including IARCs, their donors, developing country governments and research institutions, research institutions and private industry of industrialized nations, and various environmental and public advocacy concerns. Each has its own agenda and definition of what constitutes safe and appropriate release.

All biotechnology programme must abide by the legally established regulations of the country where the research is being conducted. Currently, if regulations exist at all, they vary widely from country to country. If regulations are not in place, international guidelines to ensure adequate safeguards are followed. A common first step is the successful testing of regulated organisms in industrialized countries or in countries with established regulatory and approval mechanisms; this helps clear the way for subsequent testing in developing countries. However, care must be taken when "transposing" the regulations of one country directly to another as the different ecologies, crops, quarantine regulations, etc. vary greatly from one site to another; each site must make its own policy and cost/benefit analyses (Sawyer and Dodds, 1991).

When applying new technologies, IARCs must deal with many issues including scientific, technical, environmental, regulatory and policy dimensions of research. Several centres have established special committees to prepare for the testing of such genetically improved material. IARCs follow international guidelines set by OECD, and work closely with host countries in determining proper risk analysis for field testing.

## Plant Genetic Resources and Intellectual Property Management

The need to preserve future biodiversity has been, and continues to be, a top priority for CGIAR. Over the years, several of the centres have assembled global germplasm collections for their mandated crops, often using the material to re-establish individual country collections (e.g. because of storage failures or losses caused by political unrest). The former International Board for Plant Genetic Resources (IBPGR) now the International Plant Genetic Resources Institute (IPGRI)-was established in 1974 for the purpose of promoting germplasm utilization and conservation.

*Figure: Plant Genetic Resources Project*

CGIAR has always operated an "open door" germplasm policy-meaning that all clients, whether public or private, in developed or developing countries, can enjoy access to the germplasm. But the involvement of the private sector in biotechnology research has raised some questions about both access of developing countries to new biotechnologies and the extent to which research will benefit resource-poor farmers. Developing countries see themselves as faced with two possible scenarios, either of which may be entirely acceptable. They can protect intellectual property rights, and thereby obtain access to the latest crop technologies which in turn may lead to higher seed costs. Or they can choose to forego IPR protection, perhaps to save foreign exchange and attempt to protect farmers' incomes, but risk being left behind in the technology race. (Platais and Collinson, 1992). Are alternative scenarios possible?

Intellectual property considerations are not new in CGIAR. ILRAD has long sought vaccines for its two target diseases, trypanosomiasis and theileriosis. ILRAD realized that to develop a vaccine and make it available to African farmers paradoxically it would have to file for patent protection. Such protection was considered essential to satisfy private sector partners who otherwise would be unwilling to provide the investment funds necessary to develop and produce a vaccine and bring it to market. Other centres have obtained patents on discoveries, such as small farm machines developed at IRRI, but such patents were taken for defensive purposes. CGIAR does not seek financial gain from discoveries made at IARCs. When patents are taken, they are used by IARCs themselves or transferred to their partners to ensure that scientific advances are utilized in developing countries.

In 1982 CGIAR began to review the implications of intellectual property protection, beginning with a report by TAC on the role of plant breeders' rights, and culminating in a recent discussion chapter on the subject of plant genetic resources, biosafety and intellectual property rights. The latter chapter is currently under review both within and outside the CGIAR system. At the mid-term meeting of CGIAR in Istanbul, in May 1992, the members of CGIAR agreed to a CGIAR Working Document on Genetic Resources and Intellectual Property which stated:

*"The CGIAR reaffirms that genetic resources maintained in the genebanks of the centres are held in trust for the world community. Material from the genebanks at the centres will continue to be freely available, in accordance with the 1989 CGIAR Policy on Plant Genetic Resources... Centres do not seek intellectual property protection unless it is absolutely necessary to ensure access by developing countries to new technologies and products. The centres will not seek intellectual property protection as a source of operating funds.*

*Should exceptional cases arise where a centre might receive a financial return, an appropriate means will be used to ensure that such funds are used for the conservation of genetic resources and related research" (CGIAR, 1992).*

Future CGIAR efforts in biotechnology are likely to emphasize genetic characterization of germplasm materials of important crop plants and animals, improved diagnostic tools, and enhanced genetic improvement of important crops and livestock.

Partnerships will continue to be sought with advanced institutions and developing country institutions to help solve important common problems, in particular those of developing countries.

As now, IARCs will largely be consumers and not originators of developments in basic sciences, but will stand among the leaders in developing tools and technologies that can be used in developing countries. Furthermore, IARCs will play a bridging role in technology transfer and a leading role in helping to train developing country scientists.

## THE PATTERN IN DEVELOPED COUNTRIES

In the industrialized countries, a new pattern of biotechnological research funding has emerged. With the availability of property right protection of biotechnologies and prospects of vast markets for biotechnology products and the techniques, the bulk of the research is funded, carried out, and controlled by the private sector.

For instance, since 1976 when GENENTECH, the first biotechnology company, was established, the new techniques have spawned a variety of industries that now comprise more than 400 start-up firms, more than 200 established firms that have diversified into biotechnology, and more than 200 supply firms in the United States alone.

The new biotechnology industry in the United States produced pharmaceutical, diagnostic tests and agricultural products worth close to US$2 billion in 1990. A similar trend is seen in Europe and Japan. It is estimated that about 60 percent of the biotechnology research and development funding in industrialized countries is from the private sector.

Thus, the private sector has been a major force in enhancing the capability of these countries in this field. Research institutions in the public sector are now generally required to raise a substantial part of their budget from non-government sources, via contractual research, licensing agreements and royalties.

This is tending to increase secrecy over research findings and to hinder free scientific communication. University professors, researchers and government institution scientists are increasingly becoming entrepreneurial and are entering private industry.

Another important trend is that large multinational corporations are purchasing smaller seed and biotechnology companies and diversifying their holdings.

This allows them to develop a package sale of chemicals, seeds and equipment. Heavy involvement of the private sector and market considerations greatly influence the topics and commodities chosen for research. Major crops, commodities and farming systems of great socio-economic importance to the developing world, but of little international market importance, do not figure in the biotechnology research agenda of industrialized countries. Furthermore, these countries are keen to reduce their production costs, increase the productivity, quality and value of their products and, thus, improve their overall competitiveness in the world market.

Biotechnology facilities are being established in most developing countries. However, the level of research, development and use of biotechnology for agriculture, forestry and fisheries in the developing countries is generally far below the level in the industrialized countries. Among developing countries, the status varies considerably.

A few, such as Brazil, Mexico, India, China and The Republic of Korea, have sought to gain full scientific and technological capacity, especially in agricultural biotechnology. Others, such as Indonesia, Malaysia, the Philippines and Thailand and a few countries of Latin America, have built the capacity to apply biotechniques and develop biotechnologies useful for their agriculture and food industries. The participation of the private sector in gaining biotechnological capacity is not significant in most of these countries. Many developing countries have inadequate funding, poor human resources and limited access to information, resulting in a relatively low-level capacity for research and technology development and exploitation, especially in the field of modern biotechnology research, which is costly and requires highly trained personnel. Most developing countries are vague as to their immediate aims in agricultural biotechnology. Few have appropriate proprietary-right protection systems or mechanisms to increase their access to protected techniques and products. Furthermore, there is negligible involvement of the private sector, which accentuates the problem of insufficient attention to biotechnology.

One of the major constraints on biotechnological development in developing countries is the poor quality and extent of higher education in frontier sciences, especially molecular biology. Besides, there are no or very weak links between universities and research institutions to reinforce each other and to utilize synergistically the scarce trained personnel resources. The university experts play a negligible role in national policy formulation, including that on biotechnology. Furthermore the universities are generally not oriented for efficiency in commercialization and are therefore not able to market biotechnology products of their own or others' inventions.

Some international and donor-country public sector institutions strive to ensure that attention is given to solving, through biotechnology, the food and agriculture problems of developing countries. Among this group are the International Agricultural Research Centres (IARCs) located or operating in developing countries. Capacity exists in several of these centres for utilization of advances in biotechnologies to solve certain problems of plant production and protection and of livestock production and health, and capacity is being built in others.

For example, the International Rice Research Institute (IRRI) is incorporating modern molecular biology in its rice research activities. Its programme is linked closely to the Rockefeller Foundation Biotechnology Network. The World Bank, the International Service for National Agricultural Research (ISNAR) and the Australian Government undertook a major study on the likely impact of modern biotechnology on agriculture. This study is being followed up by the participants by an expansion in the World Bank lending for biotechnology and by ISNAR's establishing an Intermediary Biotechnology Service to provide advisory services to national agricultural research systems.

## COMPROMISING THE POTENTIAL OF BIOTECHNOLOGY

All of life involves weighing risks against benefits. Through our own experience, and by observing that of others, we assess the risks of familiar activities and, sometimes almost subconsciously, we adapt to them. A child soon learns, sometimes painfully, about the high risk of touching a hot stove. Usually without a great deal of thought, we run the hazard of shark attacks at the beach. Academics and insurance company experts have been able to quantify the risks of, say, smoking a pack of cigarettes a day, commuting to work by car, or undergoing cardiac surgery. Risk is more problematic when we are confronted with unfamiliar activities or products. In the absence of sufficient experience (what scientists would call "data") to make a confident assessment of risk, we tend to become anxious and to compensate for our lack of knowledge by overestimating the risk.

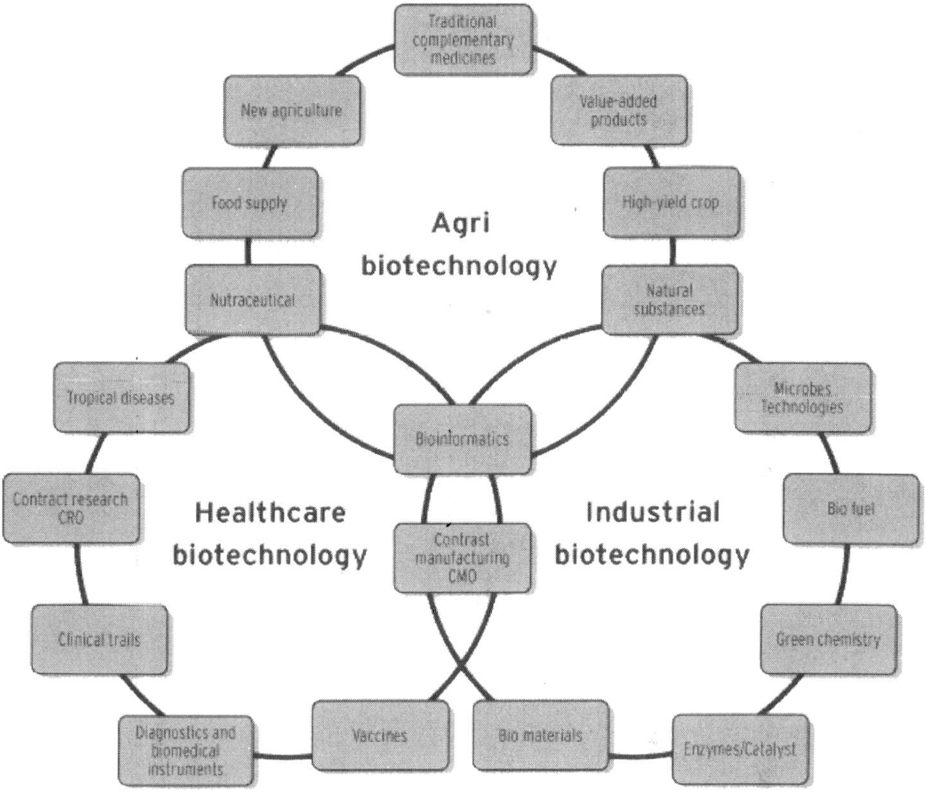


*Figure: Biotechnology focus areas.*

The authors use an apposite contemporary example to illustrate public policy run amok: the regulation in the United States and abroad of the new biotechnology, or gene-splicing, which has great potential to improve plants and microorganisms for agriculture and food production. Henry I. Miller and Gregory Conko make a persuasive case not only that the benefits of the technology far exceed its risk but also that there has been an abject failure in the formulation of public policy. The result has been, they argue correctly, gross over-regulation of the technology and its products, disincentives to research and development, and fewer choices and inflated prices for consumers.

As a plant pathologist and breeder, I have seen how the skeptics and critics of the new biotechnology wish to postpone the release of improved crop varieties in the hope that another year's, or another decade's, worth of testing will offer more data, more familiarity, more comfort. But more than a half-century in the agricultural sciences has convinced me that we should use the best that is at hand, while recognizing its imperfections and limitations. Far more often than not, this philosophy has worked, in spite of constant pessimism and scare-mongering by critics. Nevertheless, in 1963, the Rockefeller Foundation and the Mexican government formed the International Maize and Wheat Improvement Center (known by its Spanish acronym CIMMYT) and sent my team to South Asia to teach local farmers how to cultivate high-yield wheat varieties. As a result, Pakistan became self sufficient in wheat production by 1968 and India a few years later.

As we created what became known as the "Green Revolution," we confronted bureaucratic chaos, resistance from local seed breeders, and centuries of farmers' customs, habits, and superstitions. We surmounted these difficult obstacles because something new had to be done. Who knows how many would have starved if we had delayed commercializing the new high-yielding cereal varieties and improved crop management practices until we could perform tests to rule out every hypothetical problem, and test for vulnerability to every conceivable type of disease and pest? How much land for nature and wildlife habitat, and topsoil would have been lost if the more traditional, lowyield practices had not been supplanted?

At the time, Forrest Frank Hill, a Ford Foundation vice president, told me, "Enjoy this now, because nothing like it will ever happen to you again. Eventually the naysayers and the bureaucrats will choke you to death, and you won't be able to get permission for more of these efforts." Hill was right. His prediction anticipated the gene-splicing era that would arrive decades later. As Henry Miller and Gregory Conko describe in this volume, the naysayers and bureaucrats have now come into their own. If our new varieties had been subjected to the kinds of regulatory strictures and requirements that are being inflicted upon the new biotechnology, they would never have become available.

From 1950 to 1992, the world's grain output rose from 692 million tons produced on 1.70 billion acres of cropland to 1.9 billion tons on 1.73 billion acres of cropland--an increase in yield of more than 150 percent. Without high-yield agriculture, either millions would have starved or increases in food output would have been realized only through drastic expansion of acres under cultivation--with losses of pristine wilderness a hundred times greater than all the losses to urban and suburban expansion.

Today, we confront a similar problem: feeding the anticipated global population of more than eight billion people in the coming quarter of a century. The world has or will soon have the agricultural technology available to meet this challenge. The new biotechnology can help us to do things that we could not do before, and to do it in a more precise, predictable, and efficient way. The crucial question today is whether farmers and ranchers will be permitted to use that technology. Extremists in the environmental movement are doing everything they can to stop scientific progress in its tracks, and their allies in the regulatory agencies are more than eager to help. We owe a debt of gratitude to the environmental movement for raising global awareness of the importance of air and water quality, and of wildlife and wilderness preservation. It is ironic, therefore, that if the platform of anti-biotechnology extremists were to be adopted, it would have grievous consequences for both the environment and humanity. If the naysayers do manage to stop agricultural biotechnology, they might actually precipitate the famines and the crisis of global biodiversity they have been priedicting for nearly 40 years.

For a decade, the United States has produced ever-larger quantities of gene-spliced, insect-resistant corn that yields as much as or more than the best traditional hybrids but with far less need for chemical pesticides. No negative health or environmental effects have been observed. Yet there is an immensely strong, rabid anti-biotech lobby, especially in Europe, where activists have convinced many governments thwart new approvals and have opposed the use of gene-spliced corn and soybeans as food aid in famine-stricken parts of Africa and Asia. Recently, in the southern African countries of Zambia, Zimbabwe, and Angola, where many people are dying of starvation, this anti-biotech movement has helped to persuade government authorities to refuse food aid from the United States because it contains gene-spliced corn. But the risk-benefit characteristics of gene-splicing in general, and of this insect-resistant corn in particular, are extraordinarily favorable; this is an obscene exaggeration of risk.

Tragically, this is not an isolated case. There are many other examples of overreaction and resistance to technology. The American Council on Science and Health has documented a series of twenty cases--including pesticides on cranberries in 1959, the supposed hazards of cyclamates in 1969, "agent orange" in 1979, and Alar® on Pacific coast apples in 1989in which scare stories trumpeted by the media became widely known and accepted but later were shown to be of little or no consequence. The resistance to igene splicing is yet another sordid episode in this larger anti-technology, junk-science movement.

In spite of the many powerful and precise new tools and the greater health and well-being that science and technology have offered us, our society has become overly risk averse. We obsess over impurities that are detectable at levels of one part per billion, for example, while not so many years ago we would have declared a product pure if adulterants were present at less than one part per five hundred thousand. Not infrequently, regulators worry about levels of contamination not because they are worrisome but because they are detectable. They regulate not because they should but because they can. This is both foolish and destructive. Sometimes greater analytical sensitivity requires greater intellectual perspicacity.

Regulators are not alone in their demands for ever-increasing margins of safety. It seems we are born with an instinct to resist change and to regard what is new with suspicion, while forgetting the faults and risks that were readily tolerated in the past and that were eliminated or ameliorated by new technologies; chlorination of water, pasteurization, and vaccination readily come to mind. Our very wealth and well-being, made possible by technology, now seem to offer the luxury of forgoing such additional improvement. But we are hard put to equal the immortal observation of the head of the U.S. patent office at the dawning of the twentieth century who suggested that the office be closed because surely everything possible had already been invented.

## APPLICATIONS, IMPACT AND POTENTIAL

Biotechnological research and development are moving at a very fast rate. For instance, three years ago, transgenics in rice, especially in japonica rices, were considered rather difficult because of the problem of regeneration through protoplast culture. But today, a large number of transgenics both in indica and japonica types are available and are being tested at various stages. It is therefore difficult to predict the potential and impact of a given technology beyond five years or so. This section gives a brief account of the current level of technology and refers to the developments likely to occur in the next three to five years.

The potential of genetic engineering in medicine has received much attention. The potential for advances in agriculture, forestry and fishery are similarly bright. Biotechniques are already being used to create new strains of crop plants, new plant and animal diagnostic products, animal vaccines, biological pesticides and herbicides, other biological control agents, and modifications in domestic animals used for food production.

In some cases, applications are being held in check by the need for still more research to ensure that there are no harmful effects (or, if there are, they are greatly outweighed by the benefits) and by the slow pace of evolution of regulations governing the release of genetically modified organisms or products for general use or acquisition.

Biotechnological research, especially genetic engineering, on problems of field and tree crops, forest species and fisheries has a relatively short history.

The knowledge of biological aspects of species and ecosystems needs to be expanded in may cases before new biotechnologies can be applied effectively. Even where transfer of one or more genes can be achieved reliably, expression of the transferred gene(s) may not occur in the expected way. The search for and identification and cloning of useful genes, effective vector systems, methods of gene transfer and promoter mechanisms should be intensified.

A perspective on both the difficulties and intensity of the research effort suggests that diagnostic tools will speed solutions in the next five to ten years to certain disease problems in several crop, forest, animal and fish species, such as black sigatoka of banana and plantain, and virus/viroid diseases of coconut and rice. Improved tolerance to certain physical stresses, such as heat tolerance in potatoes, may also be achieved, but prospects for improved resistance of varieties and clones to most abiotic stresses are longer term.

## Livestock

Among agricultural and allied fields, animal production and health have benefited the most from biotechnology, although practical use of transgenic livestock is only a future possibility. Wide use of monoclonal antibodies for efficient diagnostics, leading to safe and specific treatments of animal diseases, is a major breakthrough. Through genetic engineering, vaccines for the prevention of viral, bacterial, and parasitic animal diseases are rendered more effective and safe. Tailored vaccines exist for pig scours, chicken bursar disease and cattle tickborne diseases. Pathogen-specific vaccines are attractive goals. Other interesting possibilities are endocrine-directed vaccines to stimulate twinning in beef cattle, immunocastration, livestock growth rate increase and vaccines that compensate for various stress-induced production losses.

Advances in genetic engineering will also facilitate the production of male-only populations of screwworms, tsetse flies, ticks and various other ectoparasites for use in the sterile release technique of control or eradication. Furthermore mammalian tissue culture may replace whole animals in the 1990s for toxicity testing of certain chemicals. The culture technique can also be exploited for studying and analyzing pesticide metabolism and for herbicide pre-screening. In vitro fertilization and embryo sexing techniques have considerably increased the use of embryo transfer techniques for cattle breeding and trade. The value of the approach will be further enhanced if embryo cloning techniques can be reliably employed. Microbial and enzymatic treatment of roughages and genetic engineering of rumen bacteria both have great potential to improve animal nutrition. Growth hormones can be produced by genetically engineered microorganisms in quantities and at the low cost necessary for widespread use to speed and increase milk and lean meat production. Biotechnological tools (embryo culture, gene cloning, etc.) may also be used for conservation of genetic resources.

## Crops

Modern biotechnologies can add greater precision and speed to plant breeding. Transgenics have already been reported in 40 crop plants, including maize, rice, soybean, cotton, rapeseed, potato, sugar beet, tomato, potato and alfalfa, but the new varieties are yet to be used commercially. Near-future opportunities for commercial exploitation include vegetables and fruits (potato, tomato, cucumber, cantaloupe and squash), followed by legumes (alfalfa) and oilseed crops (oilseed rape). A good number of the transgenics are herbicide-resistant plants whose widespread use is somewhat controversial.

Wide use is currently made of tissue culture techniques for micropropagation of elite clones and for freeing planting materials of pathogens. Monoclonal antibodies are also in use as diagnostic aids in the detection and identification of viruses and viroids. Another culture and microspore culture giving rise to haploids are being used in variety improvement to facilitate and accelerate breeding. Molecular maps and markers are being widely used to identify genes of interest to accelerate conventional breeding programmes. Efficient biological nitrogen fixation systems and strains for efficient utilization of soil nutrients are being genetically engineered. Other long-term objectives are the genetic manipulations of photosynthesis pattern and the production of hybrid seed through apomixis. A very distant possibility is providing nitrogen fixation capacity to cereals.

## Forestry

Successful regeneration through micropropagation or somatic embryogenesis has been achieved for about 100 forest species although, for most, considerable development work would be required before commercial propagation could be contemplated. In breeding programmes incorporating clonal testing, inclusion of a micromultiplication phase may facilitate more rapid deployment of superior genotypes than that afforded by sequential multiplication by cuttings. This will particularly be the case when gene mappling techniques are sufficiently advanced to permit accurate identification of superior genotypes without replicated clonal testing in the field, especially within a breeding line where substantial genetic disequilibrium exists. Furthermore, gene mapping may allow eventual identification of valuable genes, including those that contribute to quantitative traits.

As regards the application of genetic engineering, many major traits of commercial importance to forestry are under polygenic control, and much remains to be learned about the operation of the genes involved before this technique can have a major impact for these characters.

An earlier application of genetic engineering in some forestry operations may be the introduction of genes, some already available, known to confer insect of disease resistance.

Other applications of biotechnology with obvious value for forestry, but as yet not widely supported by experimental successes, include: in vitro manipulation of the maturation state, e.g. the promotion of early flowering to reduce generation intervals, in vitro selection for traits such as disease resistance and tolerance to salinity; and the use of haploid cultures.

## Fisheries

Two broad areas of biotechnology exist within the fisheries sector, namely natural products biotechnology (including mostly marine species); and aquaculture biotechnology. Commercially valuable products such as pharmaceutical compounds, pigments, oils, sterols, alginates and agarose are being extracted from micro-and macro-algae in many parts of the world. Aquatic invertebrates are currently being screened for biologically active compounds that may have, inter alia anti-carcinogenic or antiviral properties, whereas marine bacteria are currently utilized in treatments of oil spills and holding tanks in tanker ships.

Within aquaculture, induced increases in the chromosome complement (polyploidy) of commercially important species such as salmon and oysters have increased the growth and marketability of these species. Genetically engineered microbes can be used to produce fish growth hormones, which might then be used to improve feed conversion and growth rate. Synthetic reproductive hormones are produced commercially and used to regulate fecundity, breeding cycles, growth rates and sex determination in certain cultured species.

In attempts to increase desirable culture qualities such as growth rate, disease resistance, temperature tolerance and marketability, transgenic fishes containing introduced genes from other species have been produced on an experimental scale. The application of advanced biotechnology to the fisheries sector is a relatively recent practice. Therefore, continued research can be expected to reveal additional commercially viable applications and products within this sector.

## PROCESSING AND PRODUCT QUALITY AND USES

Biotechnology could have great potential for improving the quality and diversifying the uses of agricultural, forestry and fisheries products. This technology could be used for processing products by both traditional and new methods in order to:

- improve the value, safety and quality of processed and unprocessed foods, such as genetically engineered tomatoes which soften more slowly and which could be processed at lower temperatures, resulting in a higher proportion of solids to liquid in canned tomatoes;
- produce enzymes on a commercial scale, especially using the microbial system, as substitutes for enzymes isolated from animal or plant sources. For example, chymosin, the enzyme used in the manufacture of cheese, traditionally taken from the cow, can now be made from specially engineered microorganisms. Mass production of amino acids from molasses and other agricultural resources is another example. These amino acids can improve the nutritional value of food and animal feed;
- produce and use new probes for rapid detection and accurate identification of specific microbial pathogens in food, especially diagnostics for staphylococcal enterotoxins, and others; and
- enhance non-food uses of farm products and thereby increase their market value;
- develop new uses for farm products, especially those often producing surpluses, including energy production from biomass and recycling of wastes and by products. An example of such an application is found in ethanol production from cassava and sugar cane;
- produce algal cells from, for instance, Spirulina or Chlorella in mass culture for proteins;
- enhance productivity through the application of newly developed processes such as the immobilized enzyme and plant tissue culture systems which are being used to produce useful secondary metabolites. Examples are diosgenin, a female contraceptive commonly used in China and India and shikonin, a natural pigment used for lipstick in Japan.

Biotechnology is being used in a number of countries to characterize the indigenous germplasm variability for therapeutic and biologically important substances. The use of biotechnology for the removal of toxic and antinutritional factors and for improving the quality of food products is gaining momentum. These techniques include the elimination or reduction of cyanogenic compounds in cassava and the incorporation of high protein in potato and sweet potato.

## CONSERVATION AND UTILIZATION OF GENETIC RESOURCES

New biotechnologies offer not only techniques to improve the conservation of genetic resources, but also new methods to identify, clone, transfer and express genes in different organisms. This has profound implications for the utilization of genetic resources, could broaden the germplasm base from which new genetic combinations can be created, and will allow scientists to pursue their breeding efforts with greater focus and speed.

*Figure: Genetic Resources Conservation and Utilization project*

In vitro conservation and the use of techniques for germplasm exchange of certain plant species are already proving considerably more efficient than conventional methods. Furthermore, the generation of gene libraries provides a valuable adjunct to the conventional germplasm conservation methods. Molecular methods such as Restriction Fragment Length Polymorph (RFLP), isozyme and protein analyses are already being used for the characterization of genetic resources and the identification of useful genes. However, massive use of crop varieties reproduced through in vitro culture, which contain genetically identical copies of the parent, could provoke increasing genetic erosion.

## Status of Forests and Forest Genetic Resources

In terms of its social and economic impacts, forest management in the Philippines has passed through two stages and is entering a third. In stage one, large areas of forest were converted to farmland to support a growing population. In stage two, the country sought to augment agricultural development by developing an industrial economy based largely on natural resources. Forest exploitation accelerated during this stage, resulting in a logging boom. Faced with diminishing forest resources, local forest-based industries were forced to diversify or go out of business. In the third, and current, stage, concerted efforts are being made to reverse the decline in forest area and quality, and protect the country's remaining natural forests. The contribution of forestry to the Philippine economy has been eroded by massive deforestation during the past two decades. Pressure from a growing population, demands for agricultural land, urbanization and logging are among the reasons behind forest loss. Poor planning and execution of forest management programmes, weak enforcement of forest laws, political pressures, inadequate policies for forest protection and development, and the low priority historically given to forest conservation and environmental protection have exacerbated deforestation and frustrated attempts to solve forest problems.

# 5

# Biotechnology in Livestock and Fisheries Development

The gap between human demand and the availability of livestock products is huge in Africa, and imports have been increasing fast. The productivity of African breeds is very low: mature beef cattle of 4-5 years hardly weight 300 kg (as against 400 kg at one year for exotic animals); the best African cows produced 300 litre/lactation (as against 5 000 litres and more in Europe); 100 African ewes produced 50 lambs every two or three years (as against 150 lambs a year in developed countries).

Retrospectively, production policies that aimed at developing livestock through cross-breeding of exotic grade cattle with indigenous ones have generally failed. The new approach is not to make selection of local performant animals. Biotechnology offers good methods both for selection and cross-breeding schemes, essentially through Multiovulation Embryo Transfer (MOET) and the field diagnosis of pregnancy. Yet, there is wide diversity among the countries to use new technologies. For instance, even if simpler techniques such as Artificial Insemination (AI) are considered, it is seen that in East Southern Africa the technique is widely used, whereas in West and Central Africa, because of the predominance of nomadic transhumant livestock production, the technique is least developed.

MOET is well developed in few countries such as Zimbabwe and South Africa. Cattle farmers in these countries import frozen embryos from overseas. MOET is promising especially in the development of dairy cattle farms around large cities. Both AI and MOET are being used in Open Nucleus Breeding Systems (ONBS) to improve animal production. Small-scale ONBS projects are being carried out by FAO in small ruminants in Ghana and the Gambia. Moreover, by using semen from bulls, with genetic resistance to high temperature, diseases and insects common to African countries, embryos could be produced from superior cows of other regions and then implanted into surrogate mothers within the environment where improved livestock are needed.

In animal feeding, biodegradation of low-quality forage is also a promising technology. It can be associated with chemical degradation techniques (use of urea and ammonia) to improve animal feeds during dry seasons. In some of the countries, farm diagnosis of pregnancy is also being used by field workers by using a kit based on the difference of levels of progesterone in cow's milk. In the field of animal health, Sterile Insect Technology (SIT) is well known in Africa. Sterile males of glossina are being reared in Bobo-Dioulasso and Burkina Faso for field application. Successful field projects in tsetse control have been carried out in Nigeria (BICOT Project) and in Burkina Faso. Newworld screwworm was eradicated in the Libyan Arab Jamahiriya in 199091 by the use of SIT. In disease diagnosis, ELISA is being used to evaluate rinderpest immunity in vaccinated animals. Monoclonal Antibody Techniques (MAB) is not commonly used. The two techniques are very promising because they are very simple and most effective. Trials are under way to use the ELISA test in trypanosomiasis diagnosis. In vaccine production recombinant rinderpest vaccine trials done in the United States seem ready to be utilized in a mass vaccination campaign against rinderpest in Africa. There is an increasing level of activity in hybridoma work in sub-Saharan countries.

The work of ILRAD in Kenya is addressing the problem of diseases and parasites that affect cattle. Some of the common cattle diseases are East Coast Fever, tickborne diseases and trypanosomiasis. The focused approaches to researching on these diseases is expanding to yield results that will significantly benefit cattle breeders. The development of rinderpest vaccines remaining stable at ordinary temperatures was of high practical value. Biotechnology in fisheries is recent in Africa and it is being applied essentially in private fisheries sector. Work is mostly confined to hormonal treatments for sex reversal and pawning, detection of fish and shrimps disease and development of fish feed.

## RESISTANCE/TOLERANCE TO STRESSES

Biotechnologies have been particularly effective in and could have great potential for developing genotypes resistant/tolerant to biotic and abiotic stresses commonly affecting crop, livestock, forest and fish species, and can therefore contribute to reducing inputs of pesticides and to stabilizing agriculture in marginal lands and inhospitable habitats. Genetically engineered de novo synthesis of biopesticides by host species for their own defence, development of biocapsules containing genetically engineered biopesticides and virus coat protein-mediated resistance represent tremendous opportunities for pest management. It is anticipated that genetically engineered vegetable and cereal crops, resistant to certain insects and viruses or tolerant to herbicides, will be commercially available by 1993. Genetically engineered biopesticides are near approval for sale in some countries. The efficiency of insect-based control of pest arthropods, pathogens and weeds and the production of sterile males for insect control can be enhanced considerably through genetic engineering. Furthermore, biotechnologies are aiding conventional breeding in developing genotypes resistant/tolerant to high and low temperature regimes, drought and excess water conditions and saline and other problem soil conditions. A major concern about these solutions, however, is their sustainability if the improved attributes are based on one or only a few genes, as has often been the case.

## SUSTAINABLE PRODUCTION

Biotechnology may be used as a tool in the sustained production of crops, livestock, forest and fisheries by providing opportunities to:

- develop genetically broad-based varieties/breeds resistant/tolerant to biotic and abiotic stresses;
- increase efficiency in the conservation and enhanced use of natural resources, including genetic resources and systems;
- reduce use of off-farm inputs such as pesticides;
- develop populations adapted to marginal land and problem soil conditions; and
- produce genetically engineered organisms for degradation of toxic wastes and detoxification of chemical residues on produce or land.

In order to be not only environmentally friendly but also socio-economically viable and attractive, a more vigorous and focused research and development approach, especially by the public sector, is necessary. The private sector should also be encouraged to achieve the desired goal of sustainable development through the use of sound biotechniques.

## BIOTECHNOLOGY IN THE CGIAR SYSTEM

As countries all over the world confront the need to boost agricultural output, biotechnology which comprises techniques for using plants, animals and microbes to produce useful products or improve existing species-stands out as a promising tool. These techniques, ranging from simple tissue culture methods to advanced genetic and molecular manipulation of biological material, hold out the possibility of relieving some of the present tolerance of crops and animals to particular physical or environment stresses (e.g. salinity and drought), increasing the resistance of plants and animals to pests and pathogens, and increasing nutrient efficiency. Biotechnology has been defined thus: "Biotechnology is... comprised of a continuum of technologies, ranging from the long-established, and widely used technologies, which are based on the commercial use of microbes and other living organisms, through to the strategic research on genetic engineering of plants and animals" (Persley, 1990).

Exactly how much modern biotechnology can or cannot do-especially for the developing countries-is still an unanswered question. Up to this point, most biotechnology research has taken place in the industrialized countries, concentrating on application of interest to the developed world. But in recent years, developing countries have become more active in the field, as they wish to partake of research benefits. A major study in 1988-90 by the World Bank, the International Service for National Agricultural Research (ISNAR), and the Australian Government, concluded that there were many potential benefits from integrating modern biotechnology with conventional agricultural research. This could be done by focusing public sector investments on the "orphan commodities"-crops traditionally of less interest to the industrialized countries but of great importance to the vast numbers of resource-poor producers and consumers in the developing world. The Consultative Group on International Agricultural Research (CGIAR) is one research group which is tackling these challenges.

## THE CGIAR AND NEW TECHNOLOGIES

In a world where 800 million people, living mostly in rural areas, go hungry every day, food demand is set to double in the next thirty years and arable land is limited, advances in agriculture are critical if we are to reduce hunger and promote growth and development in a socially acceptable and environmentally sustainable way. CGIAR resources are directed to germplasm improvement, germplasm collection, sustainable production, policy and enhancing national agricultural research systems (NARS).

The majority of the crop research undertaken by the CGIAR Centers involves conventional plant breeding, a science-based extension of the seed selection practiced by farmers for millennia. For example, crop varieties observed to have high tolerance to disease are crossed with varieties known to have high yields. The desired outcome is a high yield, disease resistant variety. In the past this process relied on close observation and a great deal of trial and error. Today, new techniques enable us to create gene maps, discover precise information about the roles genes play and mark individual genes. As a result, plants with desirable characteristics can be bred much faster.

The starting point for crop improvement is typically a gene bank. The CGIAR gene banks hold over 530,000 samples of wild and domesticated crops in public trust. While originally the gene banks held little more information than the name of the crop variety and some basic characteristics, new technologies are enabling crop scientists to catalogue the gene bank samples-identifying the traits of each variety and identifying and marking the genes responsible for key traits. Crop scientists can now select a variety known to hold a specific characteristic, mark the gene responsible for the trait and cross it with another variety known to hold a second desirable characteristic. The offspring of these plants are then tested to determine if they hold the marked gene. Those that do are selected for further tests and eventual field planting. In this way, new knowledge dramatically increases the speed by which successful cross breeding and selection can take place.

These techniques are simply smart breeding-traditional breeding techniques made more efficient by new information about genes and new technologies. Where enhanced traditional breeding techniques have not been able to solve a specific problem, such as improving the vitamin content in rice, CGIAR supported crop scientists consider that genetically modifying organisms through genetic transformation can be a valuable option. Potential benefits and risks together with the social and environmental implications of all new technologies are rigorously assessed. The research Centers offer advice to national partners on social and environmental standards and biosafety options and all Center research takes place within national legislation and guidelines.

As transgenics could offer important options for meeting food demand and environmental challenges many scientists dedicated to reducing hunger and creating wealth among poor farmers consider such new technologies to be one part of the tool box of possible solutions. Currently about 7 percent of CGIAR research is dedicated to exploring the solutions new technologies have to offer and of that, approximately 3 percent is dedicated to the exploration of genetically modified organisms.

## GOALS AND STRATEGIES OF CGIAR

CGIAR was established in 1971 to support the works of several international agricultural research centres (IARCs). Today the CGIAR system consists of 18 IARCs, with support being received from some 40 donors. The CGIAR mission statement, as revised in 1991, states: "Through international research and related activities, and in partnership with national research systems, to contribute to sustainable improvements in the productivity of agriculture, forestry and fisheries in developing countries in ways that enhance the nutrition and well-being, especially of low-income people" (CGIAR, 1991 a).

The ultimate aims are improved nutrition and economic well-being for low-income people, including women, landless labourers and poor producers and consumers in both rural and urban areas. Research should contribute to self-reliance by increasing the purchasing power of the poor through lower costs and prices and through greater equity in income distribution.

It should also contribute to the quality of plant and animal products, to sustainability and stability in their supply, and to the prevention of environmental degradation through improved resource management. CGIAR is complex and entails a comprehensive programme of research and related activities. Its goals are:

- effective management and conservation of natural resources (i.e. Iand, water, forests and germplasm) for sustainable production;
- improved productivity of high-priority crops, livestock, trees and fish, and their integration into sustainable production systems;
- improved utilization of crop, livestock, tree and fish products in both rural and urban areas through post-harvest technology;
- progress towards equity (including gender equity) as well as improved diets, nutrition and family welfare, through better understanding of human linkages between production and consumption;
- appropriate policies for the increased productivity of crops, livestock, trees and fish, and for the sustainable use of natural resources;
- strengthened human resources and institutions for greater research capacity in developing countries' research systems.

The system's involvement in each of these areas varies greatly. In some, such as crop and livestock production, the centres play a major role. In other areas, the CG system is primarily a catalyst, stimulating and supporting research at other institutions.

And in other areas such as commodity conversion and utilization, the system integrates the work of other leading institutions into the results of the centre programmes. A combination of all of these approaches is used by the CG system in the area of biotechnology research.

It has been recommended that IARCs consider undertaking four initiatives in biotechnology to identify and transfer high-priority technologies to developing countries, explore opportunities for closer links with private industry, establish institutional biosafety committees, and establish a standing group of experts to deal with the role of biotechnology in world agriculture (Barkee and Plucknett, 1991).

## BIOTECHNOLOGY AND ITS APPLICATIONS IN AQUACULTURE AND FISHERIES

Biotechnology provides powerful tools for the sustainable development of aquaculture, fisheries, as well as the food industry. Increased public demand for seafood and decreasing natural marine habitats have encouraged scientists to study ways that biotechnology can increase the production of marine food products, and making aquaculture as a growing field of animal research. Biotechnology allows scientists to identify and combine traits in fish and shellfish to increase productivity and improve quality. Scientists are investigating genes that will increase production of natural fish growth factors as well as the natural defense compounds marine organisms use to fight microbial infections. Modern biotechnology is already making important contributions and poses significant challenges to aquaculture and fisheries development. It perceives that modern biotechnologies should be used as adjuncts to and not as substitutes for conventional technologies in solving problems, and that their application should be need-driven rather than technology-driven.

The use of modern biotechnology to enhance production of aquatic species holds great potential not only to meet demand but also to improve aquaculture. Genetic modification and biotechnology also holds tremendous potential to improve the quality and quantity of fish reared in aquaculture. There is a growing demand for aquaculture; biotechnology can help to meet this demand. As with all biotech-enhanced foods, aquaculture will be strictly regulated before approved for market. Biotech aquaculture also offers environmental benefits. When appropriately integrated with other technologies for the production of food, agricultural products and services, biotechnology can be of significant assistance in meeting the needs of an expanding and increasingly urbanized population in the next millennium. Successful development and application of biotechnology are possible only when a broad research and knowledge base in the biology, variation, breeding, agronomy, physiology, pathology, biochemistry and genetics of the manipulated organism exists. Benefits offered by the new technologies cannot be fulfilled without a continued commitment to basic research. Biotechnological programmes must be fully integrated into a research background and cannot be taken out of context if they are to succeed.

Indian fisheries and aquaculture is an important sector of food production, providing nutritional security to the food basket, contributing to the agricultural exports and engaging about fourteen million people in different activities. With diverse resources ranging from deep seas to lakes in the mountains and more than 10% of the global biodiversity in terms of fish and shellfish species, the country has shown continuous and sustained increments in fish production since independence. Constituting about 4.4% of the global fish production, the sector contributes to 1.1% of the GDP and 4.7% of the agricultural GDP.

The total fish production of 6.57 million metric tonnes presently has nearly 55% contribution from the inland sector and nearly the same from culture fisheries. Fish and fish products have presently emerged as the largest group in agricultural exports of India.( Marine products export review-MPEDA.,April 2006-March2007).The potential area of biotechnology in aquaculture include the use of synthetic hormones in induced breeding, transgenic fish,gene banking, uniparental and polyploidy population and health management.

## Biotechnology in fish breeding

Gonadotropin releasing hormone (GnRH) is now the best available biotechnological tool for the induced breeding of fish. GnRH is the key regulator and central initiator of reproductive cascade in all vertebrates.It is a decapeptide and was first isolated from pig and ship hypothalami with the ability to induce pituitary release of luteinising hormone (LH) and follicle stimulating hormone (FSH) (Schally et al.,1973).Since then only one form of GnRH has been identified in most placental mammals including human beings as the sole neuropeptide causing the release of LH and FSH.

However,in non mammalian species (except guinea pig) twelve GnRH variants have now been structurally elucidated,among them seven or eight different forms have been isolated from fish species.. The most recent GnRH purified and characterized was by Carolsfeld et al.(2000) and Robinson et al.(2000).Depending on the structural variant and their biological activities, number of chemical analogues have seen prepared and one of them is salmon GnRH analogue profusely used now in fish breeding and marked commercially under the name of Ovaprim throughout the world. The induced breeding of fish is now successfully achieved by development of GnRH technology.

*Figure: Protect Fishers*

## Transgenesis

Transgenesis or transgenics may be defined as the introduction of exogenous gene / DNA into host genome resulting in its stable maintenance, transmission and expression. The technology offers an excellent opportunity for modifying or improving the genetic traits of commercially important fishers, mollusks and crustaceans for aquaculture. The idea of producting transgenic animals became popular when Palmitter et al. (1982) first produced transgenic mouse by introducing metallothionein human growth hormone fusion gene (mT-hGH) into mouse egg, resulting in dramatic increase in growth. This triggered a series of attemptson gene transfer in economically important animals including fish.

The first transgenic fish was produced Zhu et al. (1985) in China, who claimed the transient expression n putative transgenics, although they gave no molecular evidence for the integration of the transgene. The technique has now seen successfully applied to a number of fish species. Dramatic growth enhancement has been shown using this technique especially in salmonids (Devlin et al., 1994). Some studies have revealed enhancement of growth in adult salmon to an average of 3 Đ 5 times the size of non Đ transgenic controls, with some individuals, especially during the first few months of growth, reaching as much as 10 Đ 30 times the size of the controls.The introduction of transgenic technique has simultaneously put more emphasis on the need for production of sterile progeny in order to minimize the risk of transgenic stocks mixing in the wild populations. The technical development has expanded the possibilities for producing either sterile fish or those whose reproductive activity can be specifically turned on or off using inducible promoters. This would clearly be of considerable value allowing both optimal growth and controlled reproduction of the transgenic stocks while ensuring that any escaped fish would be unable to breed.

An increased resistance of fish to cold temperatures has been another subject of research in fish transgenics for the past several years. Coldwater temperatures pose a considerable stressor to many fish and few are able to survive water temperatures much below 0-1oC. this is often a major problem in aquaculture in cold climates. Interestingly, some marine teleosts have high levels (10 Ð 25 mg/ml) of serum antifreeze proteins (AFP) or glycoproteins (AFGP) which effectively reduce the freezing temperature by preventing ice-crystal growth.

The isolation, characterization and regulation of these antifreeze proteins particularly of the inter flounder Pleuronectas americanus has been the subject of research for a considerable period in Canada. Consequently, the gene encoding the liver AFP from winter flounder was successfully introduced into the genome of Atlantic salmon where it became integrated into the germ line and then passed onto the off Ð spring F3 where it was expressed specifically in the liver (Hew et al., 1995).The introduction of AFPs to gold fish also increased their cold tolerance, to temperatures at which all the control fish died (12 h at 0o C; Wang et al., 1995). Similarly, injection or oral administration of AFP to juvenile milkfish or tilapia led to an increase in resistance to a 26 to 13o C. drop in temperature (Wu et al., 1998). The development of stocks harbouring this gene would be a major benefit in commercial aquaculture in counties where winter temperatures often border the physiological limits of these species.

The most promising tool for the future of transgenic fish production is undoubtedly in the development of the embryonic stem cell (ESC) technology. There cells are undifferentiated and remain totipotent so they can be manipulated in vitro and subsequently reintroduce into early embryos where they can contribute to the germ line of the host. This would facilitate the genes to be stably introduced or deleted (Melamed et al., 2002). Although significant progress has been made in several laboratories around the world, there are numerous problems to be resolved before the successful commercialization of the transgenic brood stock for aquaculture. To realize the full potential of the transgenic fish technology in aquaculture, several important scientific break Đ through are required. There include (i) more efficient technologies for mass gene transfer (ii) targeted gene transfer technologies such as embryonic stem cell gene transfer (iii) suitable promoters to direct the expression of transgenes at optimal levels during the desired developmental stages (iv) identified genes of desireable traits for aquaculture and other applications (v) informations on the physiological, nutritional, immunological and environmental factors that maximize the performance of the transgenics of the transgenics and (vi) safety and environmental impacts of transgenic fish.

## Chromosome Engineering

Chromosome sex manipulation techniques to induce polyploidy (triploidy and tetraploidy) and uniparental chromosome inheritance (gynogenesis and androgenesis) have been applied extensively in cultured fish species. These techniques are important in the improvement of fish breeding as they provide a rapid approach for gonadal sterilization, sex control improvement of hybrid viability and clonation. Most vertebrates are diploid meaning that they possess two complete chromosome sets in their somatic cells. Polyploidy individuals possess on or more additional chromosome sets, bringing the total to three in triploids, four in tetraploids and so on. Induced triploidy is widely accepted as the most effective method for producing sterile fish for aquaculture and fisheries management. The methods used to induce triploids and other types of chromosome set manipulations in fishes and the applications of these biotechnologies to aquaculture and fisheries management are well described. Tetraploid breeding lines are of potential benefit to aquaculture, by providing a convenient way to produce large numbers of sterile triploid fish through simple interploidy crosses between tetraploids and diploids. Although tetraploidy has been induced in many finfish species, the viability of tetraploids was low in most instances.

In teleosts, technique for inducing sterility include exogenous hormone treatment (Hunter and Donaldson, 1983) and triploidy induction (Thorgaard, 1983). The use of hormone treatements, however could be limited by governmental regulation and a lack of consumer acceptance of hormone treated fish products. Triploidy can be induced by exposing eggs to physical or chemical treatment shortly after fertilization to inhibit extrusion of the second polar body triploid fish are expected to be sterile because of the failure of homologous chromosomes to synapse correctly during the first meiotic division. Methods of triploidy induction in clued exposing fertilized eggs to temperature shock (hot or cold), hydrostatic pressure shock or chemicals such an colchicines, cytochalasin-B or nitrous oxide. Triploid can also be produced by crossing teraploids and diploids. Tetraploid induction involves fertilizing eggs with normal sperm and exposing the diploid sygote for physical or chemical treatment to suppress the first mitotic division. Gynogenesis is the process of animal development with exclusive maternal inheritance. The production of gynogenetic individuals is of particular interest to fish breeders because a high level of inbreeding can be induced in single generation. Gynogenesis may also be used to produce all Ð female populations in species with female homogamety and to reveal the sex determination mechanisms in fish. It is convenient to use all female gynogenetic progenies (instead of normal bisexual progenies) for sex inversion experiments. Methodologies combining use of induced gynogenesis with hormonal sex inversion have been developed for several aquaculture species (Gomelsky et al., 2000). Androgenesis is the process by which would have commercial application in aquaculture. It can also be used in generating homozygous lines of fish and in the recovery of lost genotypes from the crypreserved sperms. Androgenetic individuals have been produced in a few species of cyprinids, cichlids and salmonids (Bongers et al., 1994).

## Biotechnology and Fish Health Management

Disease problem area major constraint for development of aquaculture. biotechnological tools such as molecular diagnostic methods, use of vaccines and immunostimulants are gaining popularity for improving the disease resistance in fish and shellish species world over for viral diseases, avoidance of the pathogen in very important.in this context there is a need to rapid method for detection of the pathogen. Biotechnological tools such as gene probes and polymerase chain reaction (PCR) are showing great potential in this area. Gene probes and PCR based diagnostic methods have developed for a number of pathogens affecting fish and shrimp ( karunasagar,1999). In case of finfish aquaculture, number of vaccine against bacteria and viruses have been developed. Some of these have been conventional vaccines consisting of killed microorgansism but new generation of vaccine consisting of protein subunit vaccine genetically engineered organism and DNA vaccine are currently under development.

In the vertebrate system, immunization against disease is a common strategy. However the immune system of shrimp is rather poorly developed, biotechnological tools are helpful for development of molecule, which can stimulate this immune system of shrimp. Recent studies have shown that the non specific defense system can be stimulated using, microbial product such as lipopolysacharides, peptidoglycans or glucans (itami et al 1998). Among the immunostimulants known to be effective in fish glucan and levamisole enhance phagocytic activities and specific antibody responses (Sakai, 1999).

## Cryopreservation of Gametes or Gene Banking

Cryopreservation is a technique, which involve long-term preservation and storage of biological material at a very low temperature usually at-196 C,the temperature of liquid nitrogen. It is based on the principle that very low temperature tranquilize or immobilize the physiological and biochemical activities of cell, thereby making it possible to keep them viable for very long period.

The technology of cryopreservation of fish spermatozoa (milt) has been adopted for animal husbandary. The first success in preserving fish sperm at low temperature was reported by Blaxter (1953) who fertilizes Herring (Clupea herengus ) eggs with frozen thawed semen.The spermatozoa of almost all cultivable fish species has now been cryopreserved (Lakra 1993). Cryopreservation overcomes problems of male maturing before female, allow selective breeding and stock improvement and enables the conservation (Harvey,1996)One of the emerging requirement for that can be used by breeders for evolving new strains. Most of the plant varieties that has been produced are based on the gene bank collections. Aquatic gene bank however suffers from the fact that at present it is possible to cryopreserve only the male gametes of finfishes and there in no viable technique for finfish eggs and embryos. However, the recent report on the freezing of shrimp embryos. However, the recent report on the freezing of shrimps embryos by subramoniam and newton (1993) and Diwan and kandaswami (1997) look promising. Therefore, it is essential that gene banking of cultivated and cultivable aquatic species be undertaken expeditiously.

## BIOTECHNOLOGIES FOR THE FISHERY SECTOR IN DEVELOPING COUNTRIES

Biotechnology in fisheries and aquaculture represents a range of technologies that present opportunities to increase growth rate in farmed species, to improve nutrition of feeds for aquaculture, to improve fish health, to help restore and protect environments, to extend the range of aquatic species and to improve management and conservation of wild stocks. In this e-mail conference, the focus will be on genetic biotechnologies, with a brief treatment of related reproductive and gene banking technologies, and the appropriateness of their application in the fishery sector in developing countries.

It is important to note that developing countries produce more fishery products from aquaculture, inland capture fisheries and marine capture fisheries than developed countries. The coverage of the biotechnologies here is not comprehensive, but should be enough to stimulate discussion in the conference.

The vast majority of aquatic genetic resources are found in wild populations of fishes, invertebrates and aquatic plants. Fishstat, the FAO database on fishery statistics, lists 1,235 taxa of common aquatic species that are harvested by humans in major fisheries; thousands more species are taken by small-scale fishers. It also contains information on 440 species that are farmed, but just 20 of these taxa account for approximately 80% of world aquaculture production. Domestication of aquatic species has not proceeded to the same level as it has in the crop and livestock sectors. Genetic biotechnologies must be used both to assist in the further domestication of aquatic species and to help manage and conserve the genetic resources found in wild populations.

## Genetic Biotechnologies in the Fishery Sector

This Background Document provides a summary of recently developed biotechnologies that could be used, or more widely used, in the fishery sector in developing countries. Genetic biotechnologies that can be used in fisheries and in aquaculture include those that help to manage genetic resources and those for genetic improvement. For management of genetic resources, markers can be used in the identification of management units and of endangered species to assist fishery management and they can also help broodstock management in stocking programmes. These markers may be genes, proteins (i.e. the products of genes), sequences of DNA or the phenotypic expression of genes (different colours, shapes etc.).

In the 1960's, analysis of proteins revealed a wealth of genetic diversity in wild populations. Protein analysis is now relatively fast and inexpensive, but it requires tissue samples to be stored and transported frozen. DNA analysis is becoming the method of choice because of the small amount of tissue needed, the fact that the tissue can be stored dried or in alcohol, and because DNA analysis reveals much more genetic variation than protein analysis.

Several kinds of DNA markers exist, such as RFLPs, AFLPs, RAPDs and microsatellites. These, as well as other kinds of markers, can be used to analyse gene frequencies and genetic variation in and between different groups of fish. Studies carried out using these technologies in fish populations have revealed high levels of genetic variation distributed throughout the fish genome.

Genetic improvement technologies cover a range of techniques requiring different levels of expertise and resources. Chromosome-set manipulation (i.e. polyploidy induction) is an established technique to increase the number of chromosome-sets (ploidy number) in an organism. Temperature, chemical and pressure shocks applied to fish eggs can be used to produce triploid (3 chromosome-sets) individuals that have desirable culture traits. Sex-reversal and the production of single sex groups of fish is also a simple technology that combines hormone treatment and chromosome-set manipulation.

Hybridisation, i.e. the mating of genetically different groups from the same species (intra-specific hybridisation) or from different species (inter-specific hybridisation), is a simple technique that is now easy to accomplish due to our increased knowledge of reproductive biology. It can be used to combine good traits from two different species into one group of fish or to transfer a characteristic of one group to another. A problem is that breeding hybrids with hybrids results in a non-uniform and unpredictable group of fish that is generally not well suited for culture. Therefore, for hybrid production, the parent-lines must be maintained pure. The above genetic improvement techniques are considered short-term strategies, where the gains are seen in one or two generations. Selective breeding is a longer-term strategy where gains are accumulated at each generation of selection. Molecular markers may now increase the efficiency of selective breeding by facilitating the identification of quantitative trait loci (QTLs), i.e. genes that control complex characters such as growth rate and environmental tolerance and, secondly, by making it possible to use molecular markers linked to QTLs to identify desirable individuals or families.

Genetic engineering and the production of transgenic organisms is an active area of research and development in aquaculture. This is a medium-long term strategy in that development and testing of stable transgenic lines requires time. The large size and hardy nature of many fish eggs allows them to be manipulated rather easily and facilitates gene transfer by direct injection of a foreign gene or by electroporation, where an electric field assists gene transfer. In the next three sections, we will briefly discuss currently available biotechnologies in the context of fishery management, aquaculture and conservation respectively.

## Biotechnologies in Fishery Management

The role that the application of genetic principles can play in the sustainable use and conservation of living aquatic resources is being increasingly appreciated by resource managers, policy makers and the international community. Fishery management requires information on the fishery resources in order to be effective. Primary information needs include:

- an identification of the resource;
- the breeding or stock structure of the resource;
- an estimate of the size of the resource; and
- the identification of key habitat that the resource requires.

Genetic analysis of the resources can address these information needs. Gene and genotype frequencies of different markers can provide information on, inter alia, species identification, population stock structure, hybridisation and gene flow. Often, data from other sources, e.g. studies of tagged fish or of external characters of fish, cannot provide such information or are extremely difficult to collect in certain areas such as large river systems, floodplains or marine areas.

The use of protein and DNA data in fishery management requires collection of baseline (or background) genetic information. Genetic data were used to determine how sub-groups of Pacific salmon differed from each other in the Pacific Northwest. This required the analysis of hundreds of stocks of salmon but, once completed, endangered stocks were identified, levels of migration were estimated, and the contribution of different stocks to a mixed stock ocean fishery was estimated.

Protein and DNA information has been used to identify endangered species that are either inadvertently captured in wild fisheries or that are purposefully taken illegally. DNA analysis of legally sold whale meat revealed that many samples came from protected species of whale and dolphin. Species of shark are often difficult to identify because it is only the fins or flesh that are for sale; DNA analysis can be used to identify the species that provided the tissue and has the added advantage that dried tissue or less than fresh samples from markets can be studied.

## Biotechnologies in Aquaculture

Genetic biotechnologies in aquaculture focus primarily on increasing growth rate, but also include disease resistance and increased environmental tolerance. There are several biotechnologies that can be applied to farmed aquatic species. Selective breeding, i.e. traditional animal breeding, started with the common carp several thousand years ago. However, it has only recently been applied to a handful of other species of food fish such as catfish, trout and tilapia. Therefore, many farmed aquatic species are very similar to their wild relatives. Selective breeding programmes have yielded significant and consistent gains of 5-20% per generation in species of, inter alia, Atlantic salmon, catfish and tilapia.

Hybridisation is a simple genetic technology that has become easier with the development of artificial breeding techniques, such as the use of pituitary gland extract and other hormones to initiate gamete development and induce spawning (i.e. the depositing of eggs), and an increased understanding of environmental cues that influence reproduction, such as day length, temperature or water current. Many of the natural reproductive isolating mechanisms that species develop in the wild can now be overcome by fish farmers. These improvements in reproductive technologies have also assisted aquaculturists greatly in their efforts to domesticate aquatic species. In addition, by making it possible to remove the natural constraints and timing of breeding, farmers are able to mate many more species at the times that are most beneficial, and thus help to ensure a steady and consistent supply of fish to the market.

Chromosome-set manipulation can be used to produce triploid organisms that generally do not channel energy into reproduction because of problems associated with development of reproductive organs. Initially it was thought that this energy saving would result in increased growth rate, but this seems not to be the case. The real advantage of triploids seems to be in their functional sterility. For example, triploid oysters do not produce gonads (i.e. reproductive glands) and are therefore marketable at times of the year when mature oysters have an off-taste because of production of gametes (i.e. sex cells-the ovum, or egg (female), and sperm (male)). In aquaculture, one sex is often more desirable than the other. For example, female sturgeon produce caviar, male tilapia grow faster than females whereas it is the female trout and salmon that generally grow faster than the males. The production of single sex groups of fish takes advantage of these differences between the sexes and can be accomplished by manipulation of the developing gametes and embryo. The manipulation can be in the form of denaturing (i.e. destroying) the DNA in gametes followed by chromosome-set manipulation or by hormonal sex-reversal and subsequent breeding. The phenotypic sex of many aquatic species can be changed by administering appropriate hormones. For example, genetically male tilapia can be turned into females through estrogen treatments. These genetic males when mated with normal males produce a group of all-male tilapia that grow faster and have less unwanted matings (that lead to overcrowding and stunting) than a group of mixed-sex tilapia. Some of the all-male offspring would have two male chromosomes and these could be used as broodstock for subsequent generations, thus avoiding the use of hormones in the broodstock. Hybridisation can also be used to produce single sex groups of fish, when the sex-determining mechanisms in the parental lines are different (for example, hybridisation of Nile tilapia and the blue tilapia).

Genetic engineering is a vague term that has come to be nearly synonymous with gene transfer i.e. the production of transgenic fish or genetically modified organisms (GMOs). This technology is progressing rapidly and it is now possible to move genes between distantly related species. Gene transfer in fish has usually involved genes that produce growth hormone and has been shown to dramatically increase growth rate in carp, catfish, salmon, tilapia, mudloach and trout. A gene from the winter flounder that produces an anti-freeze protein was put into salmon in the hope of extending the farming range of the fish. The gene did not produce enough of the protein to extend the salmon's range into colder waters, but it did allow the salmon to continue growing during cold months when non-transgenic salmon would not grow. Transgenic technology is currently in the research and development stage. To our knowledge there are no transgenic aquatic plants or animals available to the consumer.

## Cryopreservation

The development of cryopreservation or low-temperature technology allows the short-and long-term storage of gametes. Currently, these low-temperature techniques can only be used on male gametes; eggs and embryos can generally not be stored in this way. Freezing gametes can increase the flexibility of a fish breeder, especially when breeding species where the sexes mature or migrate at different times, when the breeding season is very short, when the breeders are far apart or when one sex is exceptionally rare.

## Fish Health

Genetic biotechnologies are being used to improve fish health through conventional selection for disease resistance and through the use of molecular investigation of pathogens for characterisation and diagnosis. DNA-based technologies are being used now to characterise different species and strains of pathogens. Genetic characterisation of the pathogen may also reveal information about its origin, e.g. DNA analysis revealed two strains of crayfish plague fungus in Sweden: one from the local species and one originating in Turkey. Once the pathogen is characterised, DNA probes can be developed to screen for specific pathogens in tissue, whole animals and even in water and soil samples. These techniques are being used to detect viral diseases of marine shrimp throughout the world and for bacterial and fungal pathogens in fishes in many areas.

Genetically engineered vaccines are also being developed to protect fish against pathogens. Genetic immunisation of rainbow trout with a glycoprotein gene from the virus causing viral haemorrhagic septicaemia has recently been shown to induce high levels of protection against the virus. Work is also underway on immunising carp, salmon and other fishes with genetically engineered vaccines for other diseases.

The new molecular techniques are extremely sensitive and can identify pathogens in fish long before there are any clinical signs of the disease. This has implications for quarantine and the trade of aquatic species, which is currently governed by the World Trade Organisation and the Office International des Epizooties. Trade can be restricted based on the disease status of a product or a region; identification of minute quantities of a pathogen or of a new strain of an existing pathogen could change or influence existing trade patterns.

## Farming Systems

Farming systems for aquatic species are diverse and include industrial scale farms, family ponds and culture based fisheries (stocking), in both developed and developing countries. Often, there is a division of the production process where fingerlings (i.e. small fish, especially up to one year of age) or eggs are produced by the seed-supplier, but the grow-out to market size is done elsewhere. In the case of sea going salmon, there is often a seed supplier operating a hatchery near a river, a fingerling producer in a freshwater lake, and another group that grows the fish to market size in the sea. Marine shrimp hatcheries in Asia are usually small family owned ventures, whereas in Latin America they are more industrial in scale. Appropriateness of genetic biotechnologies must take these different systems into consideration.

## Biotechnologies in Conservation

Genetic biotechnologies can be used to reduce the impacts of farmed fish on wild populations, to identify and manage endangered species and to manage captive populations in aquaria or in species recovery programmes. In several areas, farmed fish must be made triploid, i.e. sterile, in order to reduce their impact on wild populations should they escape from the fish farm. Generally, the planned use of transgenic fish also includes the provision that they are sterile, to reduce the chance of mixing with other fishes. Genetic manipulation and polyploidisation can be combined to regenerate endangered species. This can be done from frozen sperm by denaturing the DNA in an egg of a related species, fertilising with frozen sperm from the endangered species and then duplicating the chromosome-set of the fertilised egg.

## BIOTECHNOLOGY IN SELECTED AFRICAN COUNTRIES

Summarized below is the development and application of biotechnology in ten selected African countries, representing the various subregions. The information is adopted from Bialy (1993), Chatsanga (1993), Komen and Persley (1993) and Sasson (1993).

### Burundi

Biotechnology work in Burundi was confined to in vitro culture and the promotion of biological nitrogen fixation. In vitro micropropagation was performed for rice, maize and sorghum. At the University of Burundi, Bujumbura, yam, cassava and potato were propagated in the same way. Similar work was carried out at the Agricultural Sciences Institute of Burundi. A national programme aimed at improving potato, and propagating selected varieties on a large scale, was supported by the International Potato Centre (CIP). Potato plantlets were marketed by a private company at the same time as agricultural inputs. Demand for potato plantles was high and such a marketing and distribution mechanisms was appreciated by the partners involved.

At the Agricultural and Zootechnical Institute (IRAZ) of Gitega, serving Burundi, Rwanda and Zaire, banana was micropropagated. Banana plantlets were mainly sold to Rwanda, which had a small industry for processing banana beer and wine. Since IRAZ could not meet the demand for technical reasons, Rwanda created its own micropropagation production unit and the Institute was considering giving its micropropagation work to a private company, so as to concentrate on upstream research.

### Kenya

The Kenya Agricultural Research Institute (KARI) is pioneering the development of modern biotechnological approaches to existing programmes, especially tissue culture technology for producing uniform and disease-free seedlings of potato, pyrethrum and tea. Efforts are under way to develop inoculants for leguminous plants and transfer of the technology to farmers. Until 1993 the focal point for biotechnology coordination in Kenya was the National Advisory Committee on Biotechnology Advances and their Applications (NACBAA). It was composed of the directors of research institutes under the Ministry of Research, Science and

Technology (covering agriculture, industry, health and the environment), plus representatives from the private sector. The Committee had responsibility for:

- developing a biotechnology strategy up to the year 2000;
- assessing the expected degree of scientific advancement;
- advising on modalities for handling biotechnology breakthroughs; and
- advising on the feasibility of a biotechnology park.

An outline for a national biotechnology programme was prepared in 1991 by NACBAA, identifying national priority areas, the development of regulatory guidelines, and the formation of a decentralized National Biotechnology Enterprise Programme. The programme would involve the creation of a National Biotechnology Education Centre, a Biotechnology Enhancement Fund, and a National Commission for Biotechnology to succeed NACBAA, which was disbanded after completing its task. The faculties of science and agriculture at the University of Nairobi and at Jomo Kenyatta University are also building expertise in this field, and the former intends to establish a biotechnology institute to provide a focus for biotechnology research and education.

## Mauritius

The University of Mauritius identified three areas of biotechnology research to start with, namely upgrading of sugarcane byproducts, biological nitrogen fixation and plant tissue and cell culture. In collaboration with Imperial Chemical Industries (ICI) Ltd and using ICI airlift fermenter and continuous culture techniques, molasses were used as a substrate to produce yeast single-cell protein (microbial biomass); in association with Mauritius Farms Ltd, trials were conducted on the use of this yeast biomass as a protein substitute in chicken feeds.

Investigations on the use of sugar-cane juice instead of molasses, improvement of the culture medium, selection of yeast strains, pH and temperature optimization, flocculation and sedimentation processes, yeast nutrient value and potential use as food and feed, were also planned. The digestibility of bagasse by ruminants and its nutritive value after a variety of biotechnological treatments to decrease its relative lignin content, followed by biodegradation, constituted another area of investigation. The University also foresaw the need for a databank on genetic engineering.

Several grain legume species were selected for pulse production locally to cut down on imports: Vigna mungo, Phaseolus vulgaris, Lablab purpureus, Glycine max, Lens culinaris and Cicer arietinum. Research work was being carried out on the selection of effective and persistent strains of Rhizobium for these legume species to be grown under various soil conditions and fertilizer. As regards other crops, after screening selected species of economic importance to Mauritius (tea, coffee, vanilla, fruit crops, Anthurium andreanum species and other flower species), their suitability to be cloned successfully through tissue culture was studied; these techniques were also expected to be used on a routine basis to produce selected plants with desirable attributes.

Among the requirements for biotechnology development in Mauritius, the following were identified: (i) a national programme for biotechnology development; (ii) national machinery to review advances in biotechnology applications in the Mauritian context, safety standards, training requirements, etc., taking into account assistance schemes of the United Nations system, the EEC and bilateral cooperation; (iii) a documentation centre; and (iv) funding arrangements and distribution of tasks.

Among the R&D projects, the following were considered as priority: fermentation processes, including anaerobic digestion of wastes and methane production, alcohol production, composting, degradation of lignocellulosic biomass; biofertilizers and biopesticides; virus-free crops, fruits and vegetables; production of hybrid horticultural and ornamental crops; dwarf tree crop varieties and fast-growing "energy" crops; soil-acidity tolerant crop varieties; microbial biomass for animal feed; embryo transfer in cows to increase milk yield; bovine and porcine growth hormones; diagnostic kits for common plant and animal diseases; vaccines for domestic animals (e.g. coccidiosis, Newcastle virus disease); and mass culture of micro-algae. An active effort is being made to link the public and private sectors.

## Ghana

Ghana was once the world's leading cocoa producer. By the 1970s, it ranked far behind Côte d'Ivoire, the new leader, Brazil and Malaysia. It nevertheless maintained a good reputation for cocoa, thanks to rigorous quality control. At the faculty of Agriculture of the University of Ghana, disease-resistant and higher-yielding cocoa plantlets were being multiplied through clonal propagation; thermotherapy was applied to eliminate viral and bacterial diseases. At the National Agricultural Research Institute, work was being conducted on in vitro conservation of various indigenous crop species. The Department of Veterinary Services of the University of Ghana was developing vaccines against some of the prevalent cattle diseases.

The Overseas Development Administration (ODA, United Kingdom) contributed £7.5 million of financial and technical assistance over four years to Ghana's Cocoa Research Institute and Cocoa Marketing Board. Technical assistance covered the provision of staff at the Institute and training of Ghanaians in the United Kingdom. ODA also supplied equipment for the rehabilitation of laboratories, in line with research priorities. Among other things, ODA provided assistance for the development of nucleic acid probes and Restriction fragment Lenght Polymorphism (RFLP) markers intended to understand better and prevent cocoa viral and fungal diseases. Research in these areas was conducted in collaboration with the John Innes Institute at Norwich.

## Cameroon

Micropropagation techniques have been introduced for root and tuber crops in Cameroon. Development work at the Root and Tissue Research Project Biotechnology Laboratory has focused on mass-producing disease-free seedlings from the roots and tubers of cocoyam and yam. Embryonic tissue material has been used to produce cassava seedlings. The Cameroon biotechnology capability is enhanced by other projects under way at the University of Yaoundé Biotechnology Centre which is using techniques such as cell and tissue culture, hybridoma development and recombinant DNA. The Centre's work on developing diagnostic antigens from Onchocerca volvulus, the causative agents for river blindness, is well recognized.

Bio/Afrique, a diagnostic biotechnology company in Yaoundé, was formed as a profit-making joint venture between the government of Cameroon, through the biotechnology centre at the University of Yaoundé, and private Cameroonian investors as equity partners; Applied Biotechnology (San Diego, CA) as a source of proprietary manufacturing technologies; Chiron, Crop. (Emeryville, CA), Tanox Biosystems (Houston, TX), New England Biolabs (Beverly, MA) and a number of European and American academic laboratories as scientific partners. The impact of this joint venture is yet to be felt.

## Côte d'Ivoire

The Institut international de recherche scientifique pour le développement en Afrique (IIRSDA) is the leading centre for biotechnology development in the country. Yam, a very important food crop in West Africa, is being researched in order to accelerate its rate of propagation. Studies include genome characterization, disease-free micropropagation, karyotyping, protoplast fusions and in vitro conservation. Techniques of micropropagation of cells obtained from the calli of young leaves and flowers of oil palm are being standardized. Somatic hybridization and field trials of vitro plants are being pursued. The country is the world's number one producer of cocoa and the crop accounts for almost 50 percent of export earnings. Plans to use biotechnology to develop disease-resistant and superior quality cocoa are receiving due attention.

## Nigeria

Nearly ten government research institutes of the 24 in 1993 were involved in some kind of biotechnology research, largely related to agriculture and food processing. For instance, more palatable and sweeter varieties of cassava and yam were being developed; biotechnologies were applied to improve the shelf-life of palm wine and other foods; research was being carried out on the selection of wheat lines capable of withstanding the humid conditions in southern parts of the country.

Two institutes were able to apply advanced techniques, in addition to IITA in Ibadan: the Biotechnology Unit of the Institute for Agricultural Research and Training (Ibadan), which applied tissue culture to maize (mainly for yield improvement), cowpea and groundnut; and the Biotechnology Centre of the Anambra State University of Technology, focusing on food fermentation and industrial enzyme production.

The Nigerian Institute for Oil Palm Research (Benin), the Cocoa Research Institute of Nigeria (Ibadan), and the National Root Crop Research Institute (Umudike) applied plant tissue culture and other simpler biotechniques.

The Oil Palm Institute is also working on the fermentation of oil palm sap to produce palm wine. ICRIN has also been concentrating on somatic embryogenesis, in vitro conservation of germplasm and the identification of microorganisms that could be used to improve the biochemical development of cocoa flavour during the fermentation process as well as those required for the digestion of pod husk on a large scale, to render it suitable as an animal feed.

FAO, the United Nations Development Programme (UNDP) and the International Plant Genetic Resources Institute (IPGRI) are supporting the National Centre for Genetic Resources and Biotechnology (CGRB), located at the Moor Plantation, Ibadan. It coordinates activities carried out in nine satellite laboratories, each of which has been given a mandate for research into specific biotechnology areas. Furthermore, the Ministry of Science and Technology (MOST) had set up a national committee on biotechnology and genetic engineering, with a view to pooling the scattered human resources in microbiology, genetics and molecular biology and to advising the Ministry and CGRB.

The research institutes involved in animal health were the National Livestock Research Institute, the National Animal Production and Research Institute (NAPRI), Shika, Zaria, and the National Veterinary Research Institute (NVRI), Vom, as well as the Nigerian universities at Sokoto, Ibadan, Maiduguri and Zaria, carrying out some biotechnology research. The major investigation areas included the development both of new breeds of cattle for beef and milk production, and of vaccines against a wide range of diseases.

Small-scale food processing was an important industry in Nigeria. Many products were the results of the simple but effective use of traditional biotechnologies, e.g. the production of cassava-based foodstuffs, palm wine and various other wines from local fruits such as mango, pineapple and banana, and lager beer from malted sorghum and maize.

In the early 1980s, researchers at the University of Ife identified a protein sweetener in the exceedingly sweet berry of a plant, Thaumatococcus danielli, common to the forests of southwestern Nigeria. It was subsequently confirmed that thaumatin, the sweetener in the tropical berry, was weight-for-weight at least 1 600 times sweeter than sucrose, and yet nutritious. The gene for thaumatin has since been cloned, and the protein sweetener was selling for more than US$ 5 000 per kilogram.

The Nigerian Government allocated about 45 percent of its total resources for biotechnology to agricultural biotechnologies, including those to be applied to mitigate post-harvest losses. Inadequate funding by the government and private corporations, as well as the lack of interest showen by multinational corporations was proving a major hindrance to biotechnology research and development. The total funding available for all research coordinated by the National Centre for Genetic Resources and Biotechnology was less than US$ 250 000 in 1987, while IITA was spending more than US$ 2 million on its root crop investigations alone. Another bottleneck was the shortage of trained personnel.

## Zimbabwe

The prime agency for coordinating biotechnology in Zimbabwe is the Research Council of Zimbabwe (RCZ). Its overall function is to advise the government on issues of science and technology. In its latest National Science and Technology Policy Statement, agricultural biotechnology research is a principal topic. The major trust of the biotechnology programme proposed by RCZ is to develop high-yielding crops, food technology, improved horticultural crops and improved methods of animal breeding, as well as to create a national gene bank. The proposed national programme and related issues were discussed at three workshops held in 1991.

The biotechnology programme in Zimbabwe has been spearheaded by scientists at the University of Zimbabwe which now offers an M.Sc. degree in biotechnology. The most applied early work was with the tissue culture of potato and tobacco. A private company has now taken over commercial micropropagation of potato and strawberries. Improved production of rhizobium inoculants is being contemplated.

## Senegal

Work has mostly been concentrated on biological nitrogen fixation. At the Senegalese Institute for Agricultural Research (ISRA) and the French Institute of Scientific Research for Cooperative Development (ORSTOM) laboratory in Dakar, research carried out on actinorhizian microorganisms (Frankia) aimed to identify the most effective symbiotic associations with Casuarina and Allocasuarina genotypes, micropropagated in vitro.

The identified associations could be used in reforestation schemes to be implemented in Sahelian countries, and also in North Africa and the Far East. Similar work has been done on Acacia species and their symbiotic microorganisms (rhizobium). At the same laboratory, Azolla species were studied with a view to being used as biofertilizer in paddy fields in West Africa, because of their high nitrogen-fixing capacity. Manipulation of the rhizobia of Sesbania rostrata which formed dense cover during the wet season in flooded zones was improved.

Field experiments carried out in the Casamance region using Sesbania rostrata as a green manure in rice cultivation showed that the paddy yield could be increased from 1.7 to 4 tonnes per hectare without any nitrogen fertilizer input. With a good field inoculation and good cultivation practices, this shrub legume could be used systematically and intensively in hydromorphic soils or in paddy fields.

The work should further be intensified and results shared with other ricegrowing countries in the region and outside. Other ISRA programmes focused on the biotechnological manipulation of nitrogen fixation in groundnut and Vigna unguiculata;

A West African Regional Plant Biotechnology Laboratory, through financial assistance from France, was inaugurated in 1991, at the ISRA/ORSTOM centre. A staff of about 20 Senegalese and European researchers was expected to work in the laboratory, equipped for plant tissue culture and micropropagation. This laboratory has recently been designated one of the cooperating centres of the International Centre of Genetic Engineering and Biotechnology (ICGEB).

## Zambia

The use of legume inoculants has been increasing rapidly in Zambia, demanding intensive research and development work in this field. Established in 1984, the Mt. Makulu Research Station Inoculum Production Facility has grown rapidly and promoted the application of biofertilizers over inorganic nitrogen usage.

An estimated 8 000 hectares of soybean are being inoculated annually with appropriate Rhizobium strains. Two new bioreactors were assembled and tested in 1991 at the Facility, each with a capacity to produce 100 litres of broth culture. The average annual production increase in legume inoculants on a peat-base carrier for several species has been 10-15 percent since 1987. A total of 75 000 packets, each 250 g, was supplied during 1991-1992.

## FUTURE PROSPECTS AND STRATEGIES

Alleviating the widespread food gap and attaining food security or food reliance and nutritional adequacy should be the highest national priority in almost all African countries. In order to achieve this foremost goal, production, productivity and sustainability of food and agriculture must be greatly enhanced through the generation, development and transfer of appropriate technologies. While the potential of currently available technologies should be rationally harnessed, new and emerging technologies such as biotechnologies, should be used in conjunction with, not as substitute for, existing conventional technologies.

Biotechnology is offering innovative possibilities for increasing crops, livestock, forestry and fisheries production and for protecting the environment by the reduced use of agrochemicals, conservation of genetic resources, bioremediation and recycling of wastes. In vitro culture techniques for micropropagation, freeing planting materials of viruses and other pathogens, the production of secondary metabolites and germplasm conservation are being widely used.

Genetic mapping of major tropical crops as an aid to conventional plant breeding programmes; plant virus resistance by genetic engineering of the host plant; novel biocontrol agents for pest control to reduce pesticide use; and the use of efficient and new biofertilizing agents are now being increasingly used mostly in developed and in a few developing countries. Embryo transfer techniques and the use of diagnostic kits for detection of diseases both in plants and animals, and the use of highly specific and effective vaccines are proving of great value to animal production and health. Biotechnologically designed new feeds, hormones and disease control agents are already favourably impacting fish production. But, in Africa, these technologies and products are being used only rarely, in a limited way and in an ad hoc manner.

The slow and poor development of modern biotechnology in Africa may be ascribed to: (i) lack of appropriate policies, strategies, plans and programmes and the absence of political will in general; (ii) inadequacy of human resources, paucity of financial investments, poor research and laboratory infrastructure, absence of priority-setting mechanisms and insufficiency of mechanisms to analyse global market and technology trends; and (iii) negligible involvement of the private sector and absence of effective regulatory measures.

In view of the above, the African strategy for biotechnological development should include both the provision of direct support for biotechnological research, product development, technology assessment and transfer and the creation and fostering of a climate where biotechnology can develop in a balanced manner. This will largely depend on governmental and public awareness, political will, policies and programmes for building national capabilities and strengthening/developing regional and international cooperation. Creative partnerships between public and private sector interests are critical in establishing a competitive strategy in biotechnology. While developing institutional mechanisms best suited to a given country, it is essential to establish effective interdisciplinary collaboration among molecular biologists, breeders, agronomists and plant protection specialists and to forge and menage linkages among researchers, extensionists, farmers and other users of bioproducts and techniques.

It may be generally preferable to build the new initiatives on traditional strengths in agricultural, forestry and fishery research, rather than creating totally new institutes. It should be kept in mind that biotechnology in crop and livestock production will be applied through the creation of new genotypes possessing novel characteristics, and this is possible only when conventional breeding and production systems are already in place. Development of trained personnel should be a high priority in Africa. The important areas of training that are required for a successful biotechnology programme are microbiology, plant biology, zoology, cell biology, virology, genetics, biochemistry and cell and molecular biology. In most of sub-Saharan Africa, the people trained in these areas are in very short supply. Even biology and tissue culture training is not in the currriculum of most tertiary education institutions. Very few universities offer training with adequate laboratory facilities in molecular biology and recombinant DNA technology.

Those that are trying to offer such training are constrained by the lack of equipment and staff. In recent years, however, some donor-supported formal training programmes in biotechnology have been developed. For instance, the M.Sc. in biotechnology at the University of Zimbabwe was assisted by SAREC (Sweden) and DGIS/Ministry of Foreign Affairs (the Netherlands). The donors provided the funding for both the teaching/research equipment as well as funds to buy supplies and to cover the cost of bringing Dutch and Swedish biotechnology specialists to Zimbabwe to give special lectures in areas in which there was no local competence. Students from Zimbabwe have also been admitted to universities in the donor countries for Ph.D. degrees. Such arrangements exist and/or are being developed in several other African countries. The actual and potential effects of biotechnology development are not always positive. In the African context, the negative effect is the displacement of crops at present grown for export in Africa by biotechnologically designed "industrially" produced analogous in industrialized countries. African countries should be aware of such developments and diversify their products or increase quality and cost-effectiveness of their products not only to maintain the international market share but also to enhance or open new opportunities. FAO and other concerned international systems should assist African countries in developing early warning systems whereby the potential relative substitution effects could be monitored and strategic adjustments brought about in time to avert damage.

In industrialized countries, where biotechnology is essentially in the hands of the private sector, the work is focused on areas and commodities that will fetch higher economic returns to often costly investments in biotechnology research and development. Commodities such as bananas plantations, cocoa, oil palm, roots and tubers, which are of high socio-economic importance to Africa, should be on priority biotechnology research agendas of countries in the region. International agencies and research centres should also pay due attention to "orphan" commodities.

Most of the African countries with an interest in modern biotechnology are aware of the need to institute adequate biosafety provisions and establish appropriate intellectual property and patenting systems. The GATT agreements have brought these issues to the freefront, and they must be attended to in time to enable countries to access global markets. Preliminary work on developing biosafety guidelines is underway in many African countries.

The status of the issue of intellectual property rights in the region is still in a state of flux. Policymakers and biotechnologists in the region, in consultation with concerned international agencies, need to address this issue more decisively. Access to information by establishing the necessary national and regional databases and information systems should be strengthened to render priority setting, technology assessment and transfer mechanisms more effective.

# 6

# Agriculture Safety, Food Production, Security in Green Revolution

## AGRICULTURE SAFETY AND HEALTH

Agriculture ranks among the most hazardous industries. Farmers are at high risk for fatal and nonfatal injuries, work-related lung diseases, noise-induced hearing loss, skin diseases, and certain cancers associated with chemical use and prolonged sun exposure. Farming is one of the few industries in which the families (who often share the work and live on the premises) are also at risk for injuries, illness, and death. In an average year, 516 workers die doing farm work in the U.S. (1992-2005). Of these deaths, 101 are caused by tractor overturns. Every day, about 243 agricultural workers suffer lost-work-time injuries, and about 5% of these result in permanent impairment. Agriculture is the most dangerous industry for young workers, accounting for 42% of all work-related fatalities of young workers in the U.S. between 1992 and 2000. Unlike other industries, half the young victims in agriculture were under age 15. For young agricultural workers aged 15–17, the risk of fatal injury is four times the risk for young workers in other workplaces Agricultural work exposes young workers to safety hazards such as machinery, confined spaces, work at elevations, and work around livestock.

An estimated 1.26 million children and adolescents under 20 years of age resided on farms in 2004, with about 699,000 of these youth performing work on the farms. In addition to the youth who live on farms, an additional 337,000 children and adolescents were hired to work on U.S. farms in 2004. On average, 103 children are killed annually on farms (1990-1996). Approximately 40 percent of these deaths were work-related. In 2004, an estimated 27,600 children and adolescents were injured on farms; 8,100 of these injuries were due to farm work.

## History of agriculture

Agriculture was developed at least 10,000 years ago, and it has undergone significant developments since the time of the earliest cultivation. Evidence points to the Fertile Crescent of the Middle East as the site of the earliest planned sowing and harvesting of plants that had previously been gathered in the wild. Independent development of agriculture is also believed to have occurred in northern and southern China, Africa's Sahel, New Guinea and several regions of the Americas. Agricultural practices such as irrigation, crop rotation, fertilizers, and pesticides were developed long ago but have made great strides in the past century. The Haber-Bosch method for synthesizing ammonium nitrate represented a major breakthrough and allowed crop yields to overcome previous constraints. In the past century, agriculture has been characterized by enhanced productivity, the substitution of human labour for synthetic fertilizers and pesticides, selective breeding, and mechanization. The recent history of agriculture has been closely tied with a range of political issues including water pollution, biofuels, Genetically modified organisms, tariffs, and farm subsidies. In recent years, there has been a backlash against the external environmental effects of mechanized agriculture, and increasing support for the organic movement and Sustainable agriculture.

## Timeline

Scholars have proposed a number of theories to explain the historical development of farming. Most likely, there was a gradual transition from hunter-gatherer to agricultural societies after a lengthy period during which some crops were deliberately planted and other foods were gathered in the wild. Although localised climate change is the favoured explanation for the origins of agriculture in the Levant, the fact that farming was 'invented' at least three times elsewhere, suggests that social reasons may have been instrumental. When major climate change took place after the last ice age (c.11,000 BC), much of the earth became subject to long dry seasons. These conditions favoured annual plants which die off in the long dry season, leaving a dormant seed or tuber. These plants tended to put more energy into producing seeds than into woody growth. An abundance of readily storable wild grains and pulses enabled hunter-gatherers in some areas to form the first settled villages at this time.

The Oasis Theory was proposed by Raphael Pumpelly in 1908, and popularized by Vere Gordon Childe who summarized the theory in his book

*Man Makes Himself* This theory maintains that as the climate got drier, communities contracted to oases where they were forced into close association with animals which were then domesticated together with planting of seeds. The theory has little support from contemporary scholars, as the climatex data for the time does not support the theory. The Hilly Flanks hypothesis, proposed by Robert Braidwood in 1948, suggests that agriculture began in the hilly flanks of the Taurus and Zagros mountains, and that it developed from intensive focused grain gathering in the region.

The Feasting model by Bryan Hayden suggests that agriculture was driven by ostentatious displays of power, such as throwing feasts to exert dominance. This required assembling large quantities of food which drove agricultural technology. The Demographic theories were proposed by Carl Sauer and adapted by Lewis Binford and Kent Flannery. They describe an increasingly sedentary population, expanding up to the carrying capacity of the local environment, and requiring more food than can be gathered. Various social and economic factors help drive the need for food. The evolutionary/intentionality theory, advanced by scholars including David Rindos, is the idea that agriculture is an co-evolutionary adaptation of plants and humans. Starting with domestication by protection of wild plants, followed specialisation of location and then domestication.

## Neolithic Era

Identifying an exact origin of agriculture remains problematic because the transition from hunter-gatherer societies began thousands of years before the invention of writing. Nonetheless, archaeobotanists/paleoethnobotanists have traced the selection and cultivation of specific food plant characteristics, such as a semi-tough rachis and larger seeds, to just after the Younger Dryas (about 9,500 BC) in the early Holocene in the Levant region of the Fertile Crescent. There is earlier evidence for use of wild cereals: anthropological and archaeological evidence from sites across Southwest Asia and North Africa indicate use of wild grain (e.g., from the ca. 20,000 BC site of Ohalo II in Israel, many Natufian sites in the Levant and from sites along the Nile in the 10th millennium BC). There is even evidence of planned cultivation and trait selection: grains of rye with domestic traits have been recovered from Epi-Palaeolithic (10,000+ BC) contexts at Abu Hureyra in Syria, but this appears to be a localised phenomenon resulting from cultivation of stands of wild rye, rather than a definitive step towards domestication. It isn't until after 9,500 BC that the eight so-called founder crops of agriculture appear: first emmer and einkorn wheat, then hulled barley, peas, lentils, bitter vetch, chick peas and flax. These eight crops occur more or less simultaneously on PPNB sites in the Levant, although the consensus is that wheat was the first to be sown and harvested on a significant scale.

By 7000 BC, sowing and harvesting reached Mesopotamia and there, in the fertile soil just north of the Persian Gulf, Sumerians systematized it and scaled it up. By 6000 BC farming was entrenched on the banks of the Nile River. About this time, agriculture was developed independently in the Far East, probably in China, with rice rather than wheat as the primary crop. Maize was first domesticated, probably from teosinte, in the Americas around 3000-2700 BC, though there is some archaeological evidence of a much older development. The potato, the tomato, the pepper, squash, several varieties of bean, and several other plants were also developed in the New World, as was quite extensive terracing of steep hillsides in much of Andean South America. Agriculture was also independently developed on the island of New Guinea.

In Europe, there is evidence of emmer and einkorn wheat, barley, sheep, goats and pigs that suggest a food producing economy in Greece and the Aegean by 7000 BC. Archaeological evidence from various sites on the Iberian peninsula suggest the domestication of plants and animals between 6000 and 4500 BC. Céide Fields in Ireland, consisting of extensive tracts of land enclosed by stone walls, date to 5500 BC and are the oldest known field systems in the world. The horse was domesticated in Ukraine around 4000 BC.

In China, rice and millet were domesticated by 8000 BC, followed by the beans mung, soy and azuki. In the Sahel region of Africa local rice and sorghum were domestic by 5000 BC. Local crops were domesticated independently in West Africa and possibly in New Guinea and Ethiopia. Evidence of the presence of wheat and some legumes in the 6th millennium BC have been found in the Indus Valley. Oranges were cultivated in the same millennium. The crops grown in the valley around 4000 BC were typically wheat, peas, sesame seed, barley, dates and mangoes. By 3500 BC cotton growing and cotton textiles were quite advanced in the valley. By 3000 BC farming of rice had started. Other monsoon crops of importance of the time was cane sugar. By 2500 BC, rice was an important component of the staple diet in Mohenjodaro near the Arabian Sea. By this time the Indians had large cities with well-stocked granaries. Three regions of the Americas independently domesticated corn, squashes, potato and sunflowers.

## Bronze Age

By the Bronze Age, wild food contributed a nutritionally insignificant component to the usual diet. If the operative definition of *agriculture* includes large scale intensive cultivation of land, mono-cropping, organized irrigation, and use of a specialized labour force, the title "inventors of agriculture" would fall to the Sumerians, starting ca. 5,500 BC. Intensive farming allows a much greater density of population than can be supported by hunting and gathering, and allows for the accumulation of excess product for off-season use, or to sell/barter.

The ability of farmers to feed large numbers of people whose activities have nothing to do with agriculture was the crucial factor in the rise of standing armies. Sumerian agriculture supported a substantial territorial expansion which along with internecine conflict between cities, made them the first empire builders. Not long after, the Egyptians, powered by farming in the fertile Nile valley, achieved a population density from which enough warriors could be drawn for a territorial expansion more than tripling the Sumerian empire in area.

In Sumer, barley was the primary crop; wheat, flax, dates, apples, plums, and grapes were grown as well. Mesopotamia was blessed with flooding from the Tigris and Euphrates rivers, but floods came in late spring or early summer from snow melting from the Anatolian mountains. This, along with salt deposits in the soil, made farming in Mesopotamia difficult. Sheep and goats were also domesticated, kept mainly for meat and milk, butter and cheese being made from the latter. Ur, a large town that covered about 50 acres (20 hectares), had 10,000 animals kept in sheepfolds and stables and 3,000 slaughtered every year. The city's population of 6,000 included a labour force of 2,500, cultivated 3,000 acres (12 $km^2$) of land. The labour force contained storehouse recorders, work foremen, overseers, and harvest supervisors to supplement labourers. Agricultural produce was given to temple personnel, important people in the community, and small farmers. The land was plowed by teams of oxen pulling light unwheeled plows and grain was harvested with sickles in the spring. Wagons had solid wheels covered by leather tires kept in position by copper nails and were drawn by oxen and the Syrian onager (now extinct). Animals were harnessed by collars, yokes, and headstalls. They were controlled by reins, and a ring through the nose or upper lip and a strap under the jaw. As many as four animals could pull a wagon at one time. The horse was domesticated in Ukraine around 4000 BC, and was in use by the Sumerians around 2000 BC.

## Roman Era

Roman agriculture built off techniques pioneered by the Sumerians, transmitted to them by subsequent cultures, with a specific emphasis on the cultivation of crops for trade and export. Romans laid the groundwork for the manorial economic system, involving serfdom, which flourished in the Middle Ages.

The farm sizes in Rome can be divided into three categories. Small farms were from 18-88 iugera (one iugera is equal to about 0.65 acre). Medium-sized farms were from 80-500 iugera (singular iugerum). Large estates (called latifundia) were over 500 iugera. The Romans had four systems of farm management: direct work by owner and his family; slaves doing work under supervision of slave managers; tenant farming or sharecropping in which the owner and a tenant divide up a farm's produce; and situations in which a farm was leased to a tenant. There was a great deal of commerce between the provinces of the empire, all the regions of the empire became interdependent with one another, some provinces specialized in the production of grain, others in wine and others in olive oil, depending on the soil type.

## Middle Ages

During the Middle Ages, Muslim farmers in North Africa and the Near East developed and disseminated agricultural technologies including irrigation systems based on hydraulic and hydrostatic principles, the use of machines such as norias, and the use of water raising machines, dams, and reservoirs. They also wrote location-specific farming manuals, and were instrumental in the wider adoption of crops including sugar cane, rice, citrus fruit, apricots, cotton, artichokes, aubergines, and saffron. Muslims also brought lemons, oranges, cotton, almonds, figs and sub-tropical crops such as bananas to Spain. The invention of a three field system of crop rotation during the Middle Ages, and the importation of the Chinese-invented moldboard plow, vastly improved agricultural efficiency. After 1492 the world's agricultural patterns were shuffled in the widespread exchange of plants and animals known as the Columbian Exchange. Crops and animals that were previously only known in the Old World were now transplanted to the New and vice versa. Perhaps most notably, the tomato became a favorite in European cuisine, and maize and potatoes were widely adopted. Other transplanted crops include pineapple, cocoa, and tobacco. In the other direction, several wheat strains quickly took to western hemisphere soils and became a dietary staple even for native North, Central and South Americans. Agriculture was a key element in the Atlantic slave trade, Triangular trade, and the expansion by European powers into the Americas. In the expanding Plantation economy, large plantations producing crops including sugar, cotton, and indigo, were heavily dependent upon slave labour.

## REGIONAL HISTORIES

### China

The unique tradition of Chinese agriculture has been traced to the pre-historic Xianrendong Relics and Diaotonghuan Relics (c. 12 0000 BC-7500 BC). Chinese historical and governmental records of the Warring States (481 BC-221 BC), Qin Dynasty (221 BC-207 BC), and Han Dynasty (202 BC-220 AD) eras allude to the use of complex agricultural practices, such as a nationwide granary system and widespread use of sericulture. However, the oldest extant Chinese book on agriculture is the *Chimin Yaoshu* of 535 AD, written by Jia Sixia. Although much of the literature of the time was elaborate, flowery, and allusive, Jia's writing style was very straightforward and lucid, a literary approach to agriculture that later Chinese agronomists after Jia would follow, such as Wang Zhen and his groundbreaking *Nong Shu* of 1313 AD.

Jia's book was also incredibly long, with over one hundred thousand written Chinese characters, and quoted 160 other Chinese books that were written previously (but no longer survive). The contents of Jia's 6th century book include sections on land preparation, seeding, cultivation, orchard management, forestry, and animal husbandry. The book also includes peripherally related content covering trade and culinary uses for crops.

For agricultural purposes, the Chinese had innovated the hydraulic-powered trip hammer by the 1st century BC. Although it found other purposes, its main function to pound, decorticate, and polish grain that otherwise would have been done manually. The Chinese also innovated the square-pallet chain pump by the 1st century AD, powered by a waterwheel or an oxen pulling a on a system of mechanical wheels. Although the chain pump found use in public works of providing water for urban and palatial pipe systems, it was used largely to lift water from a lower to higher elevation in filling irrigation canals and channels for farmland.

### Papua

Ancient Papuans are thought to have begun practicing agriculture around 7000 BC. They began domesticating sugarcane and root crops. Pigs may also have been domesticated around this time. By 3000 BC, Papuan agriculture was characterised by water control for irrigation.

### India

Wheat, barley, and jujube were domesticated in the Indian subcontinent by 9000 BCE; Domestication of sheep and goat soon followed. Barley and wheat cultivation—along with the domestication of cattle, primarily sheep and goat—continued in Mehrgarh culture by 8000-6000 BCE. This period also saw the first domestication of the elephant. Agro pastoralism in India included threshing, planting crops in rows—either of two or of six—and storing grain in granaries. By the 5th millennium BCE agricultural communities became widespread in Kashmir. Cotton was cultivated by the 5th millennium BCE-4th millennium BCE. Archaeological evidence indicates that rice was a part of the Indian diet by 8000 BCE. The Encyclopedia Britannica—on the subject of the first certain cultivated rice—holds that: A number of cultures have evidence of early rice cultivation, including China, India, and the civilizations of Southeast Asia.

Irrigation was developed in the Indus Valley Civilization by around 4500 BCE. The size and prosperity of the Indus civilization grew as a result of this innovation, which eventually led to more planned settlements making use of drainage and sewers. Archeological evidence of an animal-drawn plough dates back to 2500 BC in the Indus Valley Civilization.

## Mesoamerica

In Mesoamerica, the Aztecs were some of the most innovative farmers of the ancient world and farming provided the entire basis of their economy. The land around Lake Texcoco was fertile, but not large enough to produce the amount of food needed for the population of their expanding empire. The Aztecs developed irrigation systems, formed terraced hillsides, and fertilized their soil. However, their greatest agricultural technique was the chinampas, or artificial islands, also known as "floating gardens". These were used to make the swampy areas around the lake suitable for farming. To make chinampas, canals were dug through the marshy islands and shores, then mud was heaped on huge mats made of woven reeds. The mats were anchored by tying them to posts driven into the lake bed and then planting trees at their corners that took root and secured the artificial islands permanently. The Aztecs grew corn, squash, vegetables, and flowers on chinampas.

## Incas

The Andean civilizations were predominantly agricultural societies; the Incas took advantage of the ground, conquering the adversities like the Andean area and the inclemencies of the weather. The adaptation of agricultural technologies that already were used previously, allowed the Incas to organize the production a diversity of products of the coast, mountain and jungle, so them could be able to redistribute to villages that did not have access to other regions. The technological achievements reached to agricultural level, had not been possible without the workforce that was at the disposal of the Sapa Inca, as well as the road system that was allowing to store adequately the harvested resources and to distribute them for all the territory.

## Islamic World

From the 8th century, the medieval Islamic world witnessed a fundamental transformation in agriculture known as the "Muslim Agricultural Revolution" or "Arab Agricultural Revolution". Due to the trade routes established by Muslim traders across the Old World during the "Afro-Asiatic age of discovery" or "Pax Islamica", this enabled the diffusion of many crops, plants and farming techniques across the Islamic world, as well as the adaptation of crops, plants and techniques from beyond the Islamic world. These techniques included crop rotation, irrigation and pest control. Some have referred to the diffusion of numerous crops during this period as the "Globalisation of Crops",which, along with increased mechanization of agriculture, led to major changes in economy, population distribution, vegetation cover, agricultural production and income, population levels, urban growth, the distribution of the labour force, linked industries, cooking and diet, clothing, and numerous other aspects of life in the Islamic world.

Serfdom became widespread in eastern Europe in the Middle Ages. Medieval Europe owed much of its development to advances made in Islamic areas, which flourished culturally and materially while Europe and other Roman and Byzantine administered lands entered an extended period of social and economic stagnation.

As early as the ninth century, an essentially modern agricultural system became central to economic life and organization in the Arab caliphates, replacing the largely export driven Roman model. The cities of the Near East, North Africa and Moorish Spain were supported by agricultural systems which included irrigation based on knowledge of hydraulic and hydrostatic principles, some of which were continued from Roman times. In later centuries, Persian Muslims transmitted cultural elements, including advanced agricultural techniques, into Turkic lands and western India.

The Muslim agricultural revolution was based on the development of a sophisticated system of irrigation using machines such as norias, water mills, water raising machines, dams and reservoirs. With such technology they managed to greatly expand the exploitable land area. As well, they adopted a scientific approach to farming enabled them to improve farming techniques derived from the collection and collation of relevant information throughout the whole of the known world. Farming manuals were widely produced, detailing where, when, and how to plant and grow various crops. Reformed land ownership rules and labourers' rights, combining the recognition of private ownership and the rewarding of cultivators with a harvest share commensurate with their efforts also improved agricultural practices. Their counterparts in Europe struggled under a feudal system in which they were almost slaves (serfs) with little hope of improving their lot by hard work.

The introduction of new crops transforming private farming into a new global industry exported everywhere, including Europe, where farming was mostly restricted to wheat strains obtained much earlier via central Asia. Spain received what she in turn transmitted to the rest of Europe; many agricultural and fruit-growing processes, together with many new plants, fruit and vegetables. These new crops included sugar cane, rice, citrus fruit, apricots, cotton, artichokes, aubergines, and saffron. Others, previously known, were further developed. Muslims also brought to that country lemons, oranges, cotton, almonds, figs and sub-tropical crops such as bananas and sugar cane. Several were later exported from Spanish coastal areas to the Spanish colonies in the New World. Also transmitted via Muslim influence, a silk industry flourished, flax was cultivated and linen exported, and esparto grass, which grew wild in the more arid parts, was collected and turned into various articles.

## Britain

Between the 16th century and the mid-19th century, Great Britain saw a massive increase in agricultural productivity and net output. New agricultural practices like enclosure, mechanization, four-field crop rotation and selective breeding enabled an unprecedented population growth, freeing up a significant percentage of the workforce, and thereby helped drive the Industrial Revolution. By the early 1800s, agricultural practices, particularly careful selection of hardy strains and cultivars, had so improved that yield per land unit was many times that seen in the Middle Ages and before.

The 18th and 19th century also saw the development of glasshouses, or greenhouses, initially for the protection and cultivation of exotic plants imported to Europe and North America from the tropics. Experiments on Plant Hybridization in the late 1800s yielded advances in the understanding of plant genetics, and subsequently, the development of hybrid crops. Increasing dependence upon monoculture crops lead to famines and food shortages, most notably the Irish Potato Famine (1845–1849). Storage silos and grain elevators appeared in the 19th century.

## Industrialization

With the rapid rise of mechanization in the late 19th and 20th centuries, particularly in the form of the tractor, and later the Combine harvester, farming tasks could be done with a speed and on a scale previously impossible.

These advances, joined to science-driven innovations in methods and resources, have led to efficiencies enabling certain modern farms in the United States, Argentina, Israel, Germany and a few other nations to output volumes of high quality produce per land unit at what may be the practical limit.

The development of rail and highway networks and the increasing use of container shipping and refrigeration in developed nations have also been essential to the growth of mechanized agriculture, allowing for the economical long distance shipping of produce. While chemical fertilizer and pesticide have existed since the 19th century, their use grew significantly in the early twentieth century.

Until the development of chemical fertilizers, Guano was widely used as a fertilizer, though expensive. The Haber-Bosch method for synthesizing ammonium nitrate represented a major breakthrough and allowed crop yields to overcome previous constraints.

It was first patented by German chemist Fritz Haber. In 1910 Carl Bosch, while working for German chemical company BASF, successfully commercialized the process and secured further patents. Norman Borlaug and other scientists began developing crops for increased yields in the 1940s in Mexico. Their work lead to the Green Revolution, which applied western advances in fertilizer and pesticide use to farms worldwide, with varying success. Other applications of scientific research since 1950 in agriculture include gene manipulation, Hydroponics, and the development of economically viable biofuels such as Ethanol.

Though the intensive fishing practices pioneered and extended in recent history generally led to increased outputs, they have also led to the destruction of farmland, most notably in the dust bowl area of the United States following World War I. As global population increases, agriculture continues to replace natural ecosystems with monoculture crops. Since the 1970s, western farmers and consumers have become increasingly aware of, and in some cases critical of, widely used intensive agriculture practices. This growing awareness has lead to increased interest in such areas of agriculture as organic farming, permaculture, Heirloom plants and biodiversity, the growth of the Slow Food movement, and an ongoing discussion surrounding the potential for sustainable agriculture.

## AGRIBUSINESS

In agriculture, agribusiness is a generic term that refers to the various businesses involved in food production, including farming and contract farming, seed supply, agrichemicals, farm machinery, wholesale and distribution, processing, marketing, and retail sales. The term has two distinctly different connotations depending on context.

Within the agriculture industry, agribusiness is widely used simply as a convenient portmanteau of agriculture and business, referring to the range of activities and disciplines encompassed by modern food production. There are academic degrees in and departments of agribusiness, agribusiness trade associations, agribusiness publications, and so forth, worldwide. Here, the term is only descriptive, and is synonymous in the broadest sense with food industry. Among critics of large-scale, industrialized, vertically integrated food production, the term *agribusiness* is used as a negative, synonymous with corporate farming. As such, it is often contrasted with family farm. Some negative connotation is also derived from the negative associations of "business" and "corporation" from critics of capitalism or corporate excess.

An example of an agribusiness was the Old North State Winegrowers Cooperative in North Carolina. Wine grape farmers came together to not only sell their grapes but to share a winery, winemaker and marketing brand together. The cooperative failed in 2006, three years after opening its winery.

Many progressive agribusinesses are now operating online businesses. Rising fuel costs are increasingly adding financial burdens on the day to day running of agricultural companies. Agribusiness is the human activities that use natural resources to fullfill their needs of their life. Agribusiness is a broad concept used to describe corporate agricultural enterprises individually and collectively. Agribusinesses are companies involved in one or more stages of the production of crops and livestock. In other words agribusiness is an economic perspective for agricultural activities. In agriculture, agribusiness is a generic term for the various businesses involved in food production, including farming and contract farming, seed supply, agrichemicals, farm machinery, wholesale and distribution, processing, marketing, and retail sales. The term has two distinctly different connotations depending on context. Within the agriculture industry, agribusiness is widely used simply as a convenient portmanteau of agriculture and business, referring to the range of activities and disciplines encompassed by modern food production. There are academic degrees in and departments of agribusiness, agribusiness trade associations, agribusiness publications, and so forth, worldwide. Here, the term is only descriptive, and is synonymous in the broadest sense with food industry. The UN's Food and Agriculture Organization, for example, operates a section devoted to Agribusiness Development, which seeks to promote food industry growth in the Third World. Among critics of large-scale, industrialized, vertically integrated food production, the term agribusiness is used negatively, synonymous with corporate farming. As such, it is often contrasted with smaller family-owned farms. Negative connotations are also derived from the negative associations of "business" and "corporations" by critics of capitalism or corporate excess. As concern over global warming intensifies, biofuels derived from food crops quickly emerged as a practical answer to the energy crisis. Adding corn ethanol to gasoline or using palm oil for biodiesel makes the fuel burn more cleanly, stretches oil supplies, and perhaps most attractive to some politicians, provides a nice boost to big agribusiness. In Europe and in the US, increasing biofuels was mandated by law. Rising fuel costs are increasingly adding financial burdens on the day-to-day running of agricultural companies.

## GREEN REVOLUTION IN INDIA

The introduction of high-yielding varieties of seeds after 1965 and the increased use of fertilizers and irrigation are known collectively as the Green Revolution, which provided the increase in production needed to make India self-sufficient in food grains, thus improving agriculture in India. Famine in India, once accepted as inevitable, has not returned since the introduction of Green Revolution crops.

### Why Green Revolution

The world's worst recorded food disaster happened in 1943 in British-ruled India. Known as the Bengal Famine, an estimated four million people died of hunger that year alone in eastern India (that included today's Bangladesh). The initial theory put forward to 'explain' that catastrophe was that there as an acute shortfall in food production in the area. However, Indian economist Amartya Sen (recipient of the Nobel Prize for Economics, 1998) has established that while food shortage was a contributor to the problem, a more potent factor was the result of hysteria related to World War II which made food supply a low priority for the British rulers. The hysteria was further exploited by Indian traders who hoarded food in order to sell at higher prices.

Nevertheless, when the British left India four years later in 1947, India continued to be haunted by memories of the Bengal Famine. It was therefore natural that food security was a paramount item on free India's agenda. This awareness led, on one hand, to the Green Revolution in India and, on the other, legislative measures to ensure that businessmen would never again be able to hoard food for reasons of profit. However, the term "Green Revolution" is applied to the period from 1967 to 1978. Between 1947 and 1967, efforts at achieving food self-sufficiency were not entirely successful. Efforts until 1967 largely concentrated on expanding the farming areas. But starvation deaths were still being reported in the newspapers. In a perfect case of Malthusian economics, population was growing at a much faster rate than food production. This called for drastic action to increase yield. The action came in the form of the Green Revolution.

The term "Green Revolution" is a general one that is applied to successful agricultural experiments in many Third World countries. It is NOT specific to India. But it was most successful in India. What was the Green Revolution in India? There were three basic elements in the method of the Green Revolution:

(1) Continued expansion of farming areas;

(2) Double-cropping existing farmland;

(3) Using seeds with improved genetics.

## Continued expansion of farming areas

As mentioned above, the area of land under cultivation was being increased right from 1947. But this was not enough in meeting with rising demand. Other methods were required. Yet, the expansion of cultivable land also had to continue. So, the Green Revolution continued with this quantitative expansion of farmlands. However, this is NOT the most striking feature of the Revolution.

## Double-cropping existing farmland

Double-cropping was a primary feature of the Green Revolution. Instead of one crop season per year, the decision was made to have two crop seasons per year. The one-season-per-year practice was based on the fact that there is only natural monsoon per year. This was correct. So, there had to be two "monsoons" per year. One would be the natural monsoon and the other an artificial 'monsoon.' The artificial monsoon came in the form of huge irrigation facilities. Dams were built to arrest large volumes of natural monsoon water which were earlier being wasted. Simple irrigation techniques were also adopted.

## Using seeds with superior genetics

This was the scientific aspect of the Green Revolution. The Indian Council for Agricultural Research (which was established by the British in 1929 but was not known to have done any significant research) was re-organized in 1965 and then again in 1973. It developed new strains of high yield value (HYV) seeds, mainly wheat and rice but also millet and corn. The most noteworthy HYV seed was the K68 variety for wheat. The credit for developing this strain goes to Dr. M.P. Singh who is also regarded as the hero of India's Green revolution.

## Statistical Results of the Green Revolution

The Green Revolution resulted in a record grain output of 131 million tons in 1978-79. This established India as one of the world's biggest agricultural producers. No other country in the world which attempted the Green Revolution recorded such level of success. India also became an exporter of food grains around that time.

Yield per unit of farmland improved by more than 30 per cent between 1947 (when India gained political independence) and 1979 when the Green Revolution was considered to have delivered its goods.

The crop area under HYV varieties grew from seven per cent to 22 per cent of the total cultivated area during the 10 years of the Green Revolution. More than 70 per cent of the wheat crop area, 35 per cent of the rice crop area and 20 per cent of the millet and corn crop area, used the HYV seeds.

## Economic results of the Green Revolution

Crop areas under high-yield varieties needed more water, more fertilizer, more pesticides, fungicides and certain other chemicals. This spurred the growth of the local manufacturing sector. Such industrial growth created new jobs and contributed to the country's GDP.

The increase in irrigation created need for new dams to harness monsoon water. The water stored was used to create hydro-electric power. This in turn boosted industrial growth, created jobs and improved the quality of life of the people in villages.

India paid back all loans it had taken from the World Bank and its affiliates for the purpose of the Green Revolution. This improved India's creditworthiness in the eyes of the lending agencies.

Some developed countries, especially Canada, which were facing a shortage in agricultural labour, were so impressed by the results of India's Green Revolution that they asked the Indian government to supply them with farmers experienced in the methods of the Green Revolution. Many farmers from Punjab and Haryana states in northern India were thus sent to Canada where they settled (That's why Canada today has many Punjabi-speaking citizens of Indian origin). These people remitted part of their incomes to their relatives in India. This not only helped the relatives but also added, albeit modestly, to India's foreign exchange earnings.

## Sociological results of the Green Revolution

The Green Revolution created plenty of jobs not only for agricultural workers but also industrial workers by the creation of lateral facilities such as factories and hydro-electric power stations as explained above.

***Political results of the Green Revolution:*** India transformed itself from a starving nation to an exporter of food. This earned admiration for India in the comity of nations, especially in the Third World.

The Green Revolution was one factor that made Mrs. Indira Gandhi (1917-84) and her party, the Indian National Congress, a very powerful political force in India (it would however be wrong to say that it was the only reason).

## Limitations of the Green Revolution

Even today, India's agricultural output sometimes falls short of demand. The Green Revolution, howsoever impressive, has thus NOT succeeded in making India totally and permanently self-sufficient in food. In 1979 and 1987, India faced severe drought conditions due to poor monsoon; this raised questions about the whether the Green Revolution was really a long-term achievement. In 1998, India had to import onions. Last year, India imported sugar. However, in today's globalised economic scenario, 100 per cent self-sufficiency is not considered as vital a target as it was when the world political climate was more dangerous due to the Cold War.

India has failed to extend the concept of high-yield value seeds to all crops or all regions. In terms of crops, it remain largely confined to foodgrains only, not to all kinds of agricultural produce. In regional terms, only Punjab and Haryana states showed the best results of the Green Revolution. The eastern plains of the River Ganges in West Bengal state also showed reasonably good results. But results were less impressive in other parts of India. Nothing like the Bengal Famine can happen in India again. But it is disturbing to note that even today, there are places like Kalahandi (in India's eastern state of Orissa) where famine-like conditions have been existing for many years and where some starvation deaths have also been reported. Of course, this is due to reasons other than availability of food in India, but the very fact that some people are still starving in India (whatever the reason may be), brings into question whether the Green Revolution has failed in its overall social objectives though it has been a resounding success in terms of agricultural production. The Green Revolution cannot therefore be considered to be a 100 percent success.

## Initiation

The program was started with the help of the United States-based Rockefeller Foundation and was based on high-yielding varieties of wheat, rice, and other grains that had been developed in Mexico and in the Philippines. Of the high-yielding seeds, wheat produced the best results. Production of coarse grains-the staple diet of the poor and pulses the main source of protein-lagged behind, resulting in reduced per capita availability.

## Before the Green Revolution

The total area under the high-yielding-varieties program was a negligible 19,000 $km^2$ in FY 1960. Since then growth has been spectacular, increasing to nearly 154,000 $km^2$ by FY 1970, 431,000 $km^2$ by FY 1980, and 639,000 $km^2$ by FY 1990. The rate of growth decreased significantly in the late 1980s.

Nevertheless, when the British left India four years later in 1947, India continued to be haunted by memories of the Bengal Famine. It was therefore natural that food security was a paramount item on free India's agenda. This awareness led, on one hand, to the Green Revolution in India and, on the other, legislative measures to ensure that businessmen would never again be able to hoard food for reasons of profit.

## Results

The major benefits of the Green Revolution were experienced mainly in northern and northwestern India between 1965 and the early 1980s; the program resulted in a substantial increase in the production of food grains, mainly wheat and rice. Food-grain yields continued to increase throughout the 1980s, but the dramatic changes in the years between 1965 and 1980 were not duplicated. By FY 1980, almost 75 percent of the total cropped area under wheat was sown with high-yielding varieties. For rice the comparable figure was 45 percent. In the 1980s, the area under high-yielding varieties continued to increase, but the rate of growth overall was slower. The eighth plan aimed at making high-yielding varieties available to the whole country and developing more productive strains of other crops.

The Green Revolution created wide regional and interstate disparities. The plan was implemented only in areas with assured supplies of water and the means to control it, large inputs of fertilizers, and adequate farm credit.

These inputs were easily available in at least parts of the states of Punjab, Haryana, and western Uttar Pradesh; thus, yields increased most in these states. In other states, such as Andhra Pradesh and Tamil Nadu, in areas where these inputs were not assured, the results were limited or negligible, leading to considerable variation in crop yields within these states. The Green Revolution also increased income disparities: higher income growth and reduced incidence of poverty were found in the states where yields increased the most and lower income growth and little change in the incidence of poverty in other states.

## IMPACTS OF GREEN REVOLUTION IN INDIA

In the backdrop of the food crisis that gripped India in the 1960s and 1970s (and few years earlier to that too) the Government of India initiated the 'Green Revolution' program. This was an attempt to become self-sufficient in production of food grains. Traditional farming methods gave way to farming with high-yield seeds, fertilizers, and pesticides.

The Green Revolution nearly quadrupled the production of rice and wheat, transforming India's fertile areas into 'granaries'. India was no longer dependent on the foreign grain and food aid shipments from the United States. With increased production, India repaid her loans, while progressing on the path to self-sufficiency. A few decades down the road, it is evident that the benefits of the Green Revolution are associated with unanticipated harmful effects of chemicals. Numerous studies have linked the use of pesticides and chemicals to diseases such as cancer. Researchers attribute an increase in stillborn babies, and ailments such as renal failure, to the misuse of pesticides. Widespread use of pesticides has contaminated drinking water supplies and is linked to other life-threatening diseases. Source: Time Magazine.

A more painful reminder is the scene at the Bhatinda railway station in Punjab. At 9.20 pm everyday, a passenger train leaves Bhatinda town for Bikaner in Rajasthan. The train is known as the Cancer Train as it carries patients and their families to the cancer treatment center.The patients are bound for the Acharya Tulsi Regional Cancer Treatment and Research Center.

Pesticide companies blame farmers for not adhering to prescribed quantities and not using protective gear. Workers who spray the chemicals blame landlords for not investing in protection, and companies for not properly informing them of the dangers of exposure. Farmers claim it is greedy dealers who push them to spray more, and also blame the government's failure to change its policies after the harmful side effects of the Green Revolution began showing. The rampant poisoning of soil and water due to excessive use of pesticides and chemicals has left the once fertile regions barren. Unable to bear the predicament of a lost livelihood and inability to repay huge loans, farmers are committing suicides.

If India intends to avoid another food crisis in the near future, it has to address these issues now. What India needs is a farming system that is sustainable. A system that produces good yields, protects the environment, and is safe. Earlier this year, in his budget speech, Finance Minister Mr. Pranab Mukherjee announced a Rs. 400 crore scheme to take the Green Revolution to the eastern India. The practices to be followed in this second Green revolution will determine if the country learnt any lessons from the past. We cannot afford to risk the health and lives of another generation. For the farmers of Punjab and their families though, there is little respite, as they continue on their journey from Bathinda to Bikaner hoping for a cure.

## FOOD SECURITY

In a world where population growth is outstripping food supply agricultural- and especially plant-biotechnology, needs to be swiftly implemented in all walks of life. The world population is expected to reach 7 billion within 25 years, over 10 billion in the year 2050, while agricultural production is growing at the slower rate of about 1.8 percentage annually. All human beings depend on agriculture that produces food of the appropriate quality at the required quantities.

Since this can no longer be achieved by traditional methods alone, *breeding through plant biotechnology is a necessity*. In the long run the massive and immediate implementation of plant and agricultural biotechnology is more highly crucial than that of medical biotechnology, since more people worldwide die from famine and diseases related to malnutrition than from "modern", western diseases.

With "Disaster Management" becoming a central issue in modern sociology research and curricula, *food and food product shortages are the ultimate disaster*. However, unlike most natural disasters, this is one that we can prepare for and even prevent.

Domestication of plants and animals found in the wild, combined with gradual long-term changes in their qualitative and quantitative traits, were the first attributes of agriculture. Domestication, followed by food storage, coincided with the growth of microorganisms. Thus was born classical food fermentation, the earliest known application of biotechnology for the generation of food products. This traditional agriculture now faces several serious limitations:

1. Market limitations: The world is becoming a global village whose free-market rules negate the effectiveness of local pricing policies, and where a dictate of international trade and policies exists. This has greatly affected future developments in agriculture, still the world's largest business.
2. Limitations of natural resources: Global climatic changes (resulting mainly in desertification and salinization), industrialisation and urbanisation, have reduced land and water availability and caused alarming deterioration of soil, water and air quality.
3. Inherent biological genetic limitations: Although previously highly efficient, the release of new improved genotypes by classical breeding is now too slow to cope with the demands, and is considerably limited by the lack of appropriate "natural" genes that can be introgressed by traditional genetic crosses.

Only two major potential solutions seems to exist for increasing food supply and agricultural commodities, in addition to continuously improving agricultural practices, despite the aforementioned limitations:

1. A search for alternative food sources (e.g., marine or extraterrestrial products),
2. Enhanced efficient plant breeding.

Food production, for both quantity and quality, as well as for new plant commodities and products, in developed and developing countries around the globe, cannot rely solely on classical agriculture. Human survival, vis-à-vis a continuous increase in agricultural productivity, depends on the effective *merging of classical breeding with modern plant biotechnology* and the novel tools it provides.

The "green revolution", for example, increased wheat production 10-fold in India and several other countries in South East Asia, thereby feeding three times as many people. However, this revolution has already been exploited to its limits, and alternative solutions are required to breed improved crops.

Now biotechnology, integrated with classical breeding, is on the verge of creating the "evergreen revolution". The potential to improve plant and animal productivity and their proper use in agriculture relies largely on newly developed *DNA biotechnology and molecular markers.*

These techniques enable the selection of successful genotypes, better isolation and cloning of favourable traits, and the creation of transgenic organisms of importance to agriculture. Together, these generic techniques are both an extension and an integral part of classical breeding, contributing successfully to shortening breeding and selection cycles.

The new plant biotechnology implies the use of recombinant DNA techniques and in vitro cell biology in three major areas:

1. As an aid to classical breeding: This includes the ongoing genome mapping projects, e.g., in Arabidopsis, rice, maize and tomato, combined with the recent activities in functional genomics, proteomics and bioinformatics, and DNA marker-assisted selection. The combined use of these techniques will soon shorten the time required for "classical" breeding and selection cycles.
2. Generation of engineered (transgenic) organisms: In view of the inherent limitations of introgressing new genes by traditional genetic crosses (i.e., lack of appropriate desired genes and crossing barriers), the efficient engineering of plants has already resulted in improved field-grown transgenic plants in several important crops.

   The result of this impressive development, which began only 18 years ago, have made possible the direct insertion and integration of genes isolated from several organisms, and the creation of novel, and otherwise impossible genetic recombination.
3. Integration of microorganisms into plant production systems: The biotechnological development of new symbiotic, antibiotic, and antagonistic relationships between plants and microorganisms (fungi, bacteria and insects) using, among other techniques, engineered plants and microorganisms, creates new possibilities. Some of these include biological control of pests, biofertilization and plant growth stimulation, and bio- and phytoremediation.

During the last two decades, these new biotechnologies have been adapted to agricultural practices and have opened *new vistas for plant utilisation*. This will continue and intensify in the next decade. Plant biotechnology — especially in vitro regeneration and cell biology, DNA manipulation and genetic modification of biochemical pathways — is changing the plant scene in three major areas:

1. Growth and development control (vegetative, generative and propagation),
2. Protecting plants against the ever-increasing threats of abiotic and biotic stress,
3. Expanding the horizons by producing specialty foods, biochemicals and pharmaceuticals. These areas were extensively discussed at the 9th

international congress of the IAPTC and B "Plant Biotechnology and In Vitro Biology in the 21st Century", held in Jerusalem in June 1998.

## Vegetative, Generative and Propagation

The better insight into the control of plant regeneration, morphogenesis and patterns of cell division achieved during the last two decade, is due to three major discoveries:

1. The *totipotency and regeneration ability of plant cells and tissues*, as revealed by cell culture and micropropagation,
2. The elucidation of *genes responsible for hormone production and activation* in plants,
3. Active research into the mechanisms and molecular control of the *cell cycle and signal transduction pathways*, in part adopted from previous studies with animal cells, in part unique for plants. These have enabled both the *control and biotechnological manipulation* of vegetative growth, generative patterns (e.g., of flowers and seeds) and of micropropagation.

## Vegetative Growth

Morphogenetic control mechanisms are still extremely obscure, but the advent of molecular hormone and cell-cycle research is sure to lead to a better understanding of vegetative growth patterns. Thus, the possibility of biotechnologically manipulating *plant growth rate and architecture* can become a reality. For example, potential consequences of controlled auxin overproduction/availability include: adventitious root formation of importance to propagation, cell and organ elongation for biomass production, increased apical dominance of importance to timber production, etc.

Controlled cytokinin overproduction/availability can result, among other things, in enhanced bud break — which is of great importance to plant architecture, branching and compactness — a desired characteristic for some ornamentals, and delayed leaf and plant senescence.

No less important, in this respect, is the potential - as yet not practical - of affecting the orientation and rate of cell division, cell elongation and tissue longevity, by interfering with the cytoskeleton and cell cycle, the synthesis of cellulose and other cell components, and programmed cell death, respectively. A few of these possibilities have already been realised.

## Generative Development

Flowers, fruits and seeds are extremely important for agriculture. Hence, biotechnological research and development aims to interfere with and control their development and characteristics, and some of the many related studies have already produced practical applications. The major targets in *flower development* are colour, scent and senescence.

Strategies for the molecular breeding of flower colour and scent include over- and underexpression of colour (anthocyanins and carotenoids) and scent (volatiles) compounds, with respect to their biosynthesis, cellular transport and targeting. Important targets for controlling *fruit development* include growth, ripening and senescence (as for vegetative growth), colour and scent (similar to flowers) and, in addition, flavour — particularly metabolic control of sugar, acid and other flavour components.

Of great importance to fruits are biotechnological strategies for the production of seedless fruits via parthenocarpy (overproduction of auxin), pollen destruction (no fertilisation), or arrest of embryo development. The manipulation of *seed development* using biotechnological strategies is especially critical, since the seed industry (together with vegetative propagation material) constitutes the germplasm of the future for any type of plant production system.

Seeds and vegetative propagules are, practically speaking, packages of genes that form the basis of all advanced and economically viable agricultural industries, both national and private. Biotechniques and molecular strategies are now available for the major seed-based operations: hybrid seed production, generation of artificial seeds (coated somatic embryos), and for the establishment of germplasm banks that may solve some of the biodiversity issues.

## Micropropagation

Micropropagation is used routinely to generate a large number of high-quality clonal agricultural plants, including ornamental and vegetable species, and in some cases also plantation crops, fruits and vegetable species.

Micropropagation has significant advantages over traditional clonal propagation techniques.

These include the potential of combining rapid large-scale propagation of new genotypes, the use of small amounts of original germplasm (particularly at the early breeding and/or transformation stage, when only a few plants are available), and the generation of pathogen-free propagules.

This impressive application of the *principles of plant cell division and regeneration to practical plant propagation* is the result of continuous tedious studies in hundreds of labouratories worldwide, many of them in developing countries, on the standardisation of explant sources, media composition and physical state, environmental conditions and acclimatisation of in vitro plants.

Particularly noteworthy are the many recent studies on the molecular of organogenesis and somatic embryogenesis.

However, further practical applications of micropropagation, which is also commercially viable, depends on *reducing the production costs* such that it can compete with seed production or traditional vegetative propagation methods (e.g., cuttings, tubers and bulbs, grafting). Techniques that have the potential to further increase the efficiency of micropropagation, but still await further improvements, include: simplified large-scale bioreactors, cheaper automatization facilities, efficient somatic embryogenesis and synthetic seed production, greater utilisation of the autotrophic growth potential of cultures, and good repeatability and quality assurance of the micropropagated plants.

## Abiotic and Biotic Stress Tolerance

The application of molecular genetics and plant transformation to the *diagnosis and control of plant pests* has become one of the major practical success stories of plant biotechnology in the past decade. The availability of dozens of transgenic crop plants which are resistant to a range of insects, viruses and herbicides, as well as to several phytopathogenic fungi and nematodes has been validated under both field and laboratory conditions, and is of great economic importance.

Moreover, applying the principles of engineering plants for resistance to these pests to other plants of agricultural importance is now considered routine, although in practice still labourious, especially for new genotypes. Apart from a wider application to additional plants, the real *challenges lying ahead* include:

1. Improved expression of the target genes in the plants, especially their spatial and temporal control,
2. The use of wide-spectrum and alternative target genes to circumvent the problem of pest resistance,
3. Intensified integration of biological control via the use of selected and engineered microorganisms with a biocontrol potential.

While plant biotechnology has been applied successfully to fighting a large number of pests, this is not yet the case for abiotic stress conditions such as drought, salinity, extreme temperatures, chemical toxicity and oxidative stress. *Drought and salinization are the most common natural causes of lack of food and famine* in arid and semiarid regions, and the most serious environmental threats to agriculture in many parts of the world. Desertification, resulting from overexploitation by the local inhabitants, is often aggravated by regional climatic changes, and results in increased soil erosion and a decrease in land and agricultural productivity. It is estimated that increased salinization of arable land will have devastating global effects, resulting in 30 percentage land loss within the next 25 years, and up to 50 percentage in the year 2050.

Although more difficult to control and engineer than the usually monogenic traits of resistance to biotic pests and herbicides, the genetically complex response to abiotic stress is globally and regionally far more important. Therefore, *breeding for plant tolerance to drought and salinity stress* should be given a high research priority in all future agbiotech programmes.

Strategies for the manipulation of *osmotic stress tolerance* in plants might include: expression of osmoprotectants and compatible solutes, ion and water transport and channels, expression of water-binding and membrane-associated dehydrins and other proteins, transcription factors and DNA-binding proteins, etc.

Also of specific interest are the intervening stages of stress perception, signal transduction (ABA and others), and protein modification. The discovery of new stress-related genes and the design of stress-specific promoters are equally important.

## Food, Biochemicals and Pharmaceuticals

Traditionally, agriculture was targeted to improving the production of plant-derived food, in terms of both quantity and quality. This was also the initial primary target of plant biotechnology. The second phase of plant biotechnology is now gradually being implemented: a shift from the production of low-priced food and bulk commodities to high-priced, specialised plant-derived products.

This includes two major categories of biomaterials:

1. Direct improvement and modification of specialised constituents of plant origin,
2. The manufacture in plants of non-plant compounds. Biotechniques, mostly based on the *engineering of metabolic pathways*, are now available for modifying many plant constituents that are *used in the food, chemical and energy industries*. This includes many "primary" metabolites: carbohydrates (starch synthesis, yield and allocation, production of high-amylose or high-amylopectin starch, increased sucrose synthesis for the sugar industry, fructan production, etc.), proteins (improvement

of amino acid composition and protein content), oils and fats (ratio of saturated to nonsaturated fatty acids, increased content of specific valuable fatty acids like erucic acid, ricinoleic acid and others).

Many other plant constituents are either minor or non-food components, but have specific high- value applications, such as specific fatty acids as an alternative energy source, polysaccharides with heat hysteresis properties and for bioaffinity purification, temperature and salt-resistant enzymes for the food industry, etc.

Moreover, the use of plants as *"bioreactors" for the production of "foreign", non-plant compounds* is gaining momentum and may eventually lead to alternative types of agriculture. This includes, for example, production of bioactive peptides, vaccines, antibodies and a range of enzymes — mostly for the pharmaceutical industry. For the chemical industry, plants can be used to produce, e.g., polyhydroxybutyrate for the production of biodegradable thermoplastics, and cyclodextrins, which form inclusion complexes with hydrophobic substances.

Achievements today in plant biotechnology have already surpassed all previous expectations, and the future is even more promising. The full realisation and impact of the aforementioned developments depend, however, not only on continued successful and innovative research and development activities, but also on a favourable regulatory climate and public acceptance.

About 12 percentage of the world's land surface is used to grow crops, and the agricultural area required to support food production — 0.44 ha / capita in 1961 — will probably have been reduced to 0.15 ha / capita in 2050.

The intensification of agriculture with its aforementioned limitations thus requires enhanced and more efficient plant breeding and the release of economical, high-return and patentable plant-derived products. This cannot be achieved without supporting advanced research and development in biochemistry, physiology, genomics and biotechnology of agricultural plants.

Plant scientists now have a central role in society, not unlike their place 300 years ago when Jonathan Swift stated: "Whoever could make two ears of corn or two blades of grass to grow upon a spot of ground where only one grew before, would deserve better of mankind, and do more essential service to his country, than the whole race of politicians put together."

## FOOD PRODUCTION ACTUALLY RELATED TO FAMINE

To some modern Western sociologists and writers, increasing food production is not synonymous with increasing food security, and is only part of a larger equation. For example, Harvard professor Amartya Sen claimed large historic famines were not caused by decreases in food supply, but by socioeconomic dynamics and a failure of public action. However, economist Peter Bowbrick has accused Sen of misrepresenting historical data, telling outright lies and being wrong on his theory of famines. In fact Bowbrick argues that Sen's views coincide with that of the Bengal government at the time of the Bengal famine of 1943 and the policies Sen advocates failed to relieve the famine.

### *Malthusian Criticism of the Green Revolution*

Some criticisms generally involve some variation of the Malthusian principle of population. Such concerns often revolve around the idea that the Green Revolution is unsustainable, and argue that humanity is currently in a state of overpopulation with regards to the sustainable carrying capacity of the earth.

Malthusian predictions have frequently failed to materialize. In 1798 Thomas Malthus made his prediction of impending famine. The world's population had doubled by 1923 and then had doubled again by 1973 without fulfilling Malthus' prediction. Malthusian Paul R. Ehrlich, in his 1968 book *The Population Bomb,* said that India would never feed itself and claimed that "India couldn't possibly feed two hundred million more people by 1980" and "Hundreds of millions of people will starve to death in spite of any crash programs." Ehrlich's predictions failed to materialize when India became self sustaining in cereal production in 1974 (six years later) as a result of the introduction of Norman Borlaug's dwarf wheat varieties.

### **Food Production versus Quality of Diet**

Some have challenged the value of the increased food production of Green Revolution agriculture. Miguel A. Altieri, one of the pioneers of agroecology and peasant-advocate, writes that the comparison between traditional systems of agriculture and Green Revolution agriculture has been unfair, because Green Revolution agriculture produces monocultures of cereal grains, while traditional agriculture usually incorporates polycultures. This production of monoculture cereal crops is frequently used for export, feed for animals, or conversion into biofuel. According to Emile Frison from Biodiversity International, it has also lead to a change in dietary habits, as less people suffer hunger, but many are affected by malnutrition e.g. iron-deficiency or vitamin A-deficiency. Currently about 60% of annual deaths of children under the age of five are related to malnutrition. Additionally, some claim traditional systems of agriculture that were displaced by the Green Revolution such as the chinampas in Mexico or raised-field rice farming in Asia can be highly-productive. Critics point out that these traditional forms of agriculture produced less food than Green Revolution crops, and were prone to famine, as evidenced by the frequency of famine in these communities. This argument does not take into account the amount of land that was cleared or irrigated for the green revolution, and the latest statistics that show a major drop in the productivity of land that has been intensively farmed for the past 30 years due to desertification and other forms of land degradation.

There are several claims about how the Green Revolution may have decreased food security for some people. One such claim involves the shift of subsistence-oriented cropland to cropland oriented towards production of grain for export and/or animal feed. For example, the Green Revolution replaced much of the land used for pulses that fed Indian peasants for wheat, which did not make up a large portion of the peasant diet. In regards to the quality of the high yield rice (HYR) which was introduced to poverty ridden Asian countries, such as in the Philippines, which began in 1964, the flavor was found to be inferior, and was considered more glutinous and less savory than their native varieties. This resulted in a lower than average market value price. The introduction of Pesticides involved in rice production has further impacted the diets of the locals by poisoning and killing off the fish and weedy green vegetables that traditionally co-existed in the rice patties that were nutritious food sources in the diets of the Philippino farmers.

## SOCIAL AND POLITICAL IMPACTS

The Green Revolution is unpopular among many leftists because of its context within the Cold War. A major critic of the Green Revolution, the US investigative journalist Mark Dowie, writes that the primary objective of the program was a Cold War geopolitical one: providing food for the populace in underdeveloped countries which thus brought social stability and weakened the fomenting of communist insurgency. Citing internal Foundation documents, he states that the Ford Foundation had a greater concern than Rockefeller in this area.

Additionally, it is maintained elsewhere that there is a significant amount of evidence suggesting the Green Revolution had the effect of weakening socialist movements in many nations. In countries like India, Mexico, and the Philippines, technological solutions were sought as an alternative to expanding agrarian reform initiatives, the latter of which were often linked to socialist politics.

Also the water quality is being compromised. In the Aral Sea, water salinization has wiped out all native fish, leaving an economy even more dependent on the agricultural model that originated the problem.Fish are disappearing through another form of agricultural run off as well. When nitrogen intensive fertilizers are washed into waterways it results in an explosion of algae and other microorganisms that lead to oxygen depletion resulting in "dead zones", killing off fish and other creatures.

## Alternative Interpretations of the Green Revolution

Norman Borlaug has dismissed most claims of critics, but does take certain concerns seriously. He states that his work has been "a change in the right direction, but it has not transformed the world into a Utopia". Of environmental lobbyists he has stated, "some of the environmental lobbyists of the Western nations are the salt of the earth, but many of them are elitists. They've never experienced the physical sensation of hunger. They do their lobbying from comfortable office suites in Washington or Brussels. If they lived just one month amid the misery of the developing world, as I have for fifty years, they'd be crying out for tractors and fertilizer and irrigation canals and be outraged that fashionable elitists back home were trying to deny them these things".

# 7

# Development of Agriculture Biotechnology and Crop Productivity

## AGRICULTURAL BIOTECHNOLOGY SECTOR IN INDIA

Agricultural biotechnology sector in India, particularly its plant biotechnology segment, is uniquely poised for a major process of transformation. As the country's agricultural biotechnology sector made strides in the 1990s in r-DNA, transgenics and molecular marker assisted plant breeding process, the Government of India responded with a matching policy support and regulatory framework that was designed to render the path of progress in R&D, sustainable and bio-safe. To a large extent, developments in the policy front have been induced by a vibrant non-governmental sector that intensely intervened on the sensitivities of modern biotechnology.

This chapter discusses analysis of the issues and proceeds to further consider the path of progress achieved by agricultural biotechnology companies in India in the area of commercialization of biotechnology products. The special focus of this chapter is on plant biotechnology. The biotechnology sector in India in 2002-03 is estimated to have attracted private investments to the tune of US$ 140 million. Infrastructure and R&D account for nearly 80 per cent of these investments.

During the period from 1986 to 2002 it is estimated that the Government of India invested an amount of US$ 275 million in the biotechnology sector. The Government of India has also extended venture capital support for biotechnology start-ups through the Technology Development Board and the New Millennium Indian Technology Industry Leadership Initiative (NMITLI). Meanwhile the States of Andhra Pradesh, Karnataka, Maharashtra, Gujarat,

Kerala and Tamil Nadu have drawn up ambitious sub-national biotechnology policies, which promise fiscal and infrastructure support measures to prospective entrepreneurs. Some State Governments have launched their own venture funds for supporting novel start-up ventures.

To catalyse the process, the Government of India accorded priority to IPR protection. The Indian Patents Act of 1970 was amended to provide for product patents for agro-chemicals, drugs and pharmaceuticals and micro-organisms over a transitory period. A product patent regime would be in place in India by January 1, 2005. The new patent law extends protection term for inventions from 14 to 20 years. To provide for protection of IPRs in respect of new plant varieties, Parliament has enacted a *sui-generis* legislation in the shape of the Plant Varieties Protection and Farmers Rights (PVFR) Act in 2001.

These developments have created favourable legal conditions for international partnerships in biotechnology R&D. By prioritizing FDI investments in biotechnology, the Government of India has sent appropriate signals to international investors. Further the fact that these changes have been in consonance with the WTO-TRIPs has lent greater creditability to these changes. Meanwhile, the Government has also brought in regulations over agricultural biotechnology. Apart from the biosafety regulations that form part of the Environment (Protection) Act 1986, the Indian Parliament has legislated the National Biodiversity Act in the year 2002, to provide for regulatory controls over access to biological resources in India.

## AGRICULTURE IN INDIA

India's record of progress in agriculture over the past four decades has been quite impressive. The agriculture sector has been successful in keeping pace with rising demand for food. The contribution of increased land area under agricultural production has declined over time and increases in production in the past two decades have been almost entirely due to increased productivity.

Contribution of agricultural growth to overall progress has been widespread. Increased productivity has helped to feed the poor, enhanced farm income and provided opportunities for both direct and indirect employment.

The success of India's agriculture is attributed to a series of steps that led to availability of farm technologies which brought about dramatic increases in productivity in 70s and 80s often described as the Green Revolution era. The major sources of agricultural growth during this period were the spread of modern crop varieties, intensification of input use and investments leading to expansion in the irrigated area. In areas where 'Green Revolution' technologies had major impact, growth has now slowed.

New technologies are needed to push out yield frontiers, utilize inputs more efficiently and diversify to more sustainable and higher value cropping patterns.

At the same time there is urgency to better exploit potential of rainfed and other less endowed areas if we are to meet targets of agricultural growth and poverty alleviation. Given the wide range of agroecological setting and producers, Indian agriculture is faced with a great diversity of needs, opportunities and prospects.

Future growth needs to be more rapid, more widely distributed and better targeted. These challenges have profound implications for the way farmers' problems are conceived, researched and transferred to the farmers. On the one hand agricultural research will increasingly be required to address location specific problems facing the communities on the other the systems will have to position themselves in an increasingly competitive environment to generate and adopt cutting edge technologies to bear upon the solutions facing a vast majority of resource poor farmers. In the past agriculture has played and will continue to play a dominant role in the growth of Indian economy in the foreseeable future. It represents the largest sector producing around 28 percent of the GDP, is the largest employer providing more than 60 percent of the jobs and is the prime arbiter of living standards for seventy percent of India's population living in the rural areas.

These factors together with a strong determination to achieve self-sufficiency in food grains production have ensured a high priority for agriculture sector in the successive development plans of the country.

An important facet of progress in agriculture is its success in eradication of its critical dependence on imported foodgrains. In the 1950's nearly 5 percent of the total foodgrains available in the country were imported.

This dependence worsened during the 1960's when two severe drought years led to a sharp increase in import of foodgrains. During 1966 India had to import more than 10 million tonnes of foodgrains as against a domestic production of 72 million tonnes. In the following year again, nearly twelve million tonnes had to be imported. On the average well over seven percent of the total availability of foodgrains during the 1960s had to be imported.

Indian agriculture has progressed a long way from an era of frequent droughts and vulnerability to food shortages to becoming a significant exporter of agricultural commodities. This has been possible due to persistent efforts at harnessing the potential of land and water resources for agricultural purposes. Indian agriculture, which grew at the rate of about 1 percent per annum during the fifty years before independence, has grown at the rate of about 3 percent per annum in the post independence era.

## Agriculture – sub-sectors

Indian agriculture broadly consists of four sub-sectors. Agriculture proper including all food-crops oilseeds, fiber, plantation crops, fruits and vegetables is the largest accounting for nearly 70 percent of the agriculture sector as a whole.

*Table. Agriculture sub-sectors – value of output (%)*

| Sub-sector | 1960-61 | 1970-71 | 1980-81 | 1993-94 |
|---|---|---|---|---|
| Agriculture | 78.0 | 79.3 | 75.3 | 68.5 |
| Livestock | 16.6 | 15.1 | 17.2 | 24.5 |
| Forestry | 4.4 | 4.4 | 5.9 | 3.8 |
| Fisheries | 1.0 | 1.2 | 1.6 | 3.2 |
| | 100.0 | 100.0 | 100.0 | 100.0 |
| Total value at Current price (₹ Billion) | 89 | 218 | 615 | 2740 |

The rapid growth in this sub-sector through exploitation of wastelands and fallows, spread of irrigation and adoption of production enhancing technologies was critical in transforming India from a country vulnerable to food shortages to one of exportable surplus. Although this sub-sector has made impressive progress its share in the sector as a whole has declined from 78 percent in 1960-61 to less than 70 percent by early 90s.

Correspondingly the share of livestock sector has increased considerably. The livestock industry has grown from ₹ 15 billion in early 1960s to ₹ 100 billion by 1980-81 and ₹ 672 billion by 1993-94. In nominal terms the sector grew at almost 15 percent per annum during 1980s. Milk production, which was almost stagnant for two decades ending 1970, grew by over 5 percent per annum in the 80s. Similarly, production of eggs increased at the rate of about 6.5 percent during the same period. As a result the share of livestock increased from about 17 percent till early 80s to 25 percent by 1993-94.

Though it plays relatively a minor role within the sector as a whole, fishing sub-sector activities have been on the rise. The sub-sector has grown from only ₹ 3 billion in 1970-71 to nearly ₹ 90 billion in 1993-94. The growth was particularly rapid in 70s and 80s. Value added increased at over 5 percent per annum during this period.

In real terms forestry and logging activities have been on the decline since mid seventies. As of 1993-94, the size of the industry in terms of value of output was 103 billion.

*Table. Foodgrains Production*

| Year | Area (million ha) | Production (million tonnes) | Yield kg/ha |
|---|---|---|---|
| 1950-51 | 97.32 | 50.82 | 522 |
| 1960-61 | 115.58 | 82.02 | 709 |
| 1970-71 | 124.32 | 108.42 | 872 |
| 1980-81 | 126.67 | 129.59 | 1023 |
| 1990-91 | 127.84 | 176.39 | 1379 |
| 1998-99 | 124.0 | 195.25 | 1574 |

Over the past three decades, the country has successfully transformed itself from a food deficit economy to one which is essentially self sufficient in availability of foodgrains and other essential commodities, albeit only at the prevailing level of effective demand. Annual aggregate foodgrains production, which averaged about 82 million tonnes in 1960-61 increased to 123.7 and 172.5 million tonnes for the trienniums ending 1980-81 and 1990-91 respectively. Current (1998-99) production level is 195 million tonnes and the country has been able to accumulate substantial, (35 million tonnes) stocks of foodgrains to cope up with any sudden difficulties arising from drought or a similar situation in any part of the country.

Increased outputs, have been achieved chiefly by adopting, since mid sixties, a strategy aimed at increasing foodgrains production by concentrating public sector efforts and resources in regions with a high potential for quick and substantial productivity gains through increased cropping intensity and average yields. These were the areas favoured by agroclimatic resource conditions and where irrigation facilities already existed or could be developed relatively rapidly. The main elements of this strategy were:

- Expansion of irrigation coverage,
- Increased provision and utilization of key inputs – mainly high yielding varieties (HYVs) of crops, mainly of wheat and rice and chemical fertilizers and plant protection chemicals,
- Expansion and improvement of institutional support services such as research and extension and
- Price policies favourable to producers of major foodgrains.

The success of this strategy was made possible by development and availability of replicable production technology packages, so called 'Green

Revolution' technologies. Irrigation facilitated double cropping and widespread adoption of HYVs. The HYVs performed particularly well under irrigated conditions, were highly responsive to fertilizers and their short duration permitted increases in cropping intensities.

Irrigation development was the cornerstone of the strategy. Undivided India was amongst the largest irrigated areas in the world. With partition nearly one-third of the irrigated area went to Pakistan. At the time of independence the net irrigated area was 20.9 million ha (gross irrigated area 22.6 million ha).

Recognizing large-scale development of irrigation facilities as critical to rapid agricultural growth, the country has spent about ₹ 45,000 crores on irrigation development in the first four decades after independence. During the period 1950-51 to 1965-66 development of irrigation through government canals grew from 7.2 million ha to 9.8 million ha – a growth rate of 2.1 percent per annum.

During 1970s this pace dropped slightly to 1.9 percent. In 1980s the rate of increase dropped significantly to 1.1 percent per annum.

*Table. India – irrigated area*

| Year | Net | Gross | Net as % of net sown | Gross as % of gross cropped |
|---|---|---|---|---|
| | --- million ha --- | ---- % ---- | | |
| 1970-71 | 31.10 | 38.19 | 22.17 | 23.04 |
| 1980-81 | 38.72 | 49.78 | 27.66 | 28.83 |
| 1990-91 | 48.02 | 63.20 | 33.61 | 34.03 |
| 1995-96 | 53.51 | 71.51 | 37.63 | 38.33 |

The growth of tube-well irrigation, however, increased rapidly from 4.5 million ha in 1970-71 to 9.5 million ha in 1980-81 and then to 14.3 million ha by 1990-91. The net irrigated area increased from 31 million ha in 1970-71 to 53.5 million ha in 1995-96 which corresponds to 22 percent of the net sown area in 1970-71 and 37.63 percent in 1995-96 *(Tables 3, 4)*. With improvements in irrigation efficiency the gross irrigated areas has increased to 71.51 million ha. The percentage of gross cropped area service by irrigation increased from 18.3 percent in 1960-61 to 23.0 percent in 1970-71 and to over 38 percent at present.

Fertilizers have constituted yet another key input in addition to expanded irrigation and spread of HYVs in achieving goals of high production and productivity. India currently occupies third position in the world, after China and USA, in terms of fertilizer production and consumption. Consumption of fertilizers has increased from 1.54 million tonnes in 1967-68, representing the pre green revolution era, to 17.31 million tonnes currently (1997-98).

*Table. Sources of irrigation*

| Year | Net Irrigated Area (million ha) | Sources | | | | |
|---|---|---|---|---|---|---|
| | | Canals | Tanks | tube-wells | Other wells | Other Sources |
| | | (% net irrigated area) | | | | |
| 1960-61 | 24.80 | 10.4 | 4.6 | 0.2 | 7.2 | 2.4 |
| 1970-71 | 31.10 | 41.28 | 13.22 | 14.34 | 23.88 | 7.29 |
| 1980-81 | 38.72 | 39.49 | 8.22 | 24.62 | 21.08 | 6.59 |
| 1990-91 | 48.02 | 36.34 | 6.13 | 29.69 | 21.73 | 6.11 |
| 1995-96 | 53.01 | 32.04 | 5.81 | 33.52 | 22.16 | 6.47 |

The average per hectare use of fertilizers currently around 85 kg per hectares is the lowest among several Asian countries.

However, rice and wheat account for a major fraction, around 65 percent of the total fertilizer consumed in the country, with very little fertilizers going to the rainfed areas. According to some current projections, fertilizer's use will need to increase to 30-35 million tonnes to meet the foodgrains need of 2020. The demand for nutrients will stretch by almost another 15 million tonnes if requirements for horticulture, vegetables and plantation and commercial crops are included. At present domestic production of N and P fertilizers (13.42 million tonnes) falls short of consumption by over 20 percent. In addition the entire requirement of K fertilizer is imported.

Most agricultural development programmes initiated in 1960s were concentrated in regions of high potential. Thus five states, Punjab, Haryana, Uttar Pradesh, Andhra Pradesh and Tamil Nadu account for 50 percent of the country's net irrigated and 53 percent of the gross irrigated area. The combination of expanding irrigation coverage and widespread adoption of short duration HYVs led to significant increases in cropping intensities. Acreage cropped more than once per year increased from 13 million ha in 1950-51 to about 44 million ha at present. Average cropping intensity for the country as a whole rose from 115 percent in 1960-61 to 131 in 1993-94. By 1993-94 cropping intensity has risen to 187 percent in Punjab, 167 percent in Haryana and 142 percent in Uttar Pradesh.

An important consequence of the strategies adopted since sixties has been to boost production of, chiefly, two crops rice and wheat. Their share in total foodgrains production went up from 57 percent in 1970-71 to more than 75 percent in 1990-91.

*Table. Cropping intensity*

| Years | Net sown area | Gross cropped area | Area cropped more than once | Cropping intensity |
|---|---|---|---|---|
| | million ha | | | (%) |
| 1950-51 | 118.8 | 131.9 | 13.1 | 111.1 |
| 1960-61 | 133.2 | 152.8 | 19.6 | 114.3 |
| 1970-71 | 140.3 | 165.8 | 25.5 | 118.2 |
| 1980-81 | 140.0 | 172.6 | 32.6 | 123.3 |
| 1990-91 | 143.0 | 185.7 | 42.7 | 129.9 |
| 1993-94 | 142.0 | 186.4 | 44.4 | 131.2 |

Production of foodgrains other than rice and wheat did not increase significantly and in the eastern region even the yield of rice did not increase.

Agricultural production and income rose substantially in the north-western states of Punjab, Haryana, Western Uttar Pradesh, parts of Rajesthan, Tamil Nadu and Andhra Pradesh.

By contrast productivity and output growth have been modest in eastern and central India and in deccan plateau. Progress was particularly slow in rainfed areas, which account for over 60 percent of the cropped area and where a great majority of rural poor are concentrated. An important impact of the strategies pursued in the 'Green Revolution' period has been intensification of regional disparities and imbalances in agricultural development and food availability and hence levels of food security. For the country as a whole while per capita availability of cereals has increased substantially, that of pulses has decreased significantly.

In summary an annual increase in foodgrains production of 3.22 percent during fifties was mainly because of expansion in area. Sixties recorded a low annual growth rate of 1.72 percent necessitating large-scale imports of foodgrains.

*Table. Rice and Wheat Production*

*Production (million tonnes)*

| Year | Rice | Wheat | Foodgrains | R+W as % of Foodgrains |
|---|---|---|---|---|
| | million tonnes | | | |
| 1970-71 | 40.80 | 20.86 | 108.4 | 57.4 |
| 1980-81 | 49.91 | 34.55 | 129.6 | 65.4 |
| 1990-91 | 72.78 | 53.03 | 176.4 | 75.0 |
| 1996-97 | 81.74 | 69.35 | 199.4 | 75.0 |

Annual growth of 2.08 percent was recorded during seventies. This decade was the turning point in India's foodgrains economy leading to self-sufficiency through significant productivity increase first in wheat and later in rice in the eighties.

An annual growth of 3.5 percent in foodgrains in eighties was the hallmark of green revolution that enabled India to become self-sufficient and even a marginal exporter.

The pace of growth slowed in nineties barely making or even slower than the population growth rate. This is a matter of concern.

*Table. Annual compound growth rate of foodgrains production*

| Crop | 1950-51 to 1959-60 | 1960-61 to 1969-70 | 1970-71 to 1979-80 | 1980-81 to 1989-90 | 1990-91 to 1997-98 |
|---|---|---|---|---|---|
| Rice | 3.28 | - 8.05 | 1.91 | 4.29 | 1.53 |
| Wheat | 4.51 | 5.90 | 4.69 | 4.24 | 3.67 |
| Coarse cereals | 2.75 | 1.48 | 0.74 | 0.74 | - 0.49 |
| Total cereals | 3.00 | 2.51 | 2.37 | 3.63 | 1.84 |
| Pulses | 2.72 | 1.35 | - 0.54 | 2.78 | 0.76 |
| Total foodgrains | 3.22 | 1.72 | 2.08 | 3.54 | 1.66 |

*Table. Per capita net availability of cereals and pulses per day (g)*

| *Year* | *Cereals* | *Pulses* | *Total* |
|---|---|---|---|
| 1951 | 334 | 61 | 395 |
| 1961 | 400 | 69 | 469 |
| 1971 | 418 | 51 | 469 |
| 1981 | 417 | 38 | 455 |
| 1991 | 468 | 42 | 510 |
| 1998 | 451 | 33 | 484 |

## Horticulture

The diversity of physiographic, climate and soil characteristics enables India to grow a large variety of horticultural crops – fruits, vegetables, flowers, spices, aromatic and medicinal plants, plantation crops etc. India is the largest producer of fruits in the world and second largest producer of vegetables.

The area under fruits estimated at 1.45 million ha in 1970-71 grew to 2.8 million ha in 1991-92 and then more rapidly to 5 to 6 million ha by 1994-95. This sector is likely to grow rapidly in the future both on account of internal demands and export opportunities.

## Animal Husbandry and Fisheries

Animal husbandry and dairying sub-sector plays an important role in overall economy and in social development. The contribution of the sub-sector is estimated to be about 25 percent of the total value of output of agricultural sector.

*Table. Production of milk, eggs and fish in India*

| Year | Milk production / availability million tonnes/annum | g./day | Fish (000 tonnes) | Eggs (million) |
|---|---|---|---|---|
| 1950-51 | 17.0 | 124 | 752 | 1832 |
| 1960-61 | 20.0 | 124 | 1160 | 2881 |
| 1970-71 | 22.0 | 112 | 1756 | 6172 |
| 1980-81 | 31.6 | 128 | 2442 | 10060 |
| 1990-91 | 53.9 | 176 | 3836 | 21101 |
| 1998-99 | 75.0 | 210 | 5388 | 28400 |

The sector also plays a significant role in supplementing family incomes and generating employment in the rural sector particularly among the land-less, small and marginal farmers and women besides providing nutritious food. Production of suitable cross breeds and their wider adoptions has contributed to increasing country's milk production, which has now reached 75 million tonnes annually. Similarly, genetic improvement and better management practices have markedly pushed up production of poultry and eggs.

*Table. Number of holdings (million)*

| Category of Farmers | 1970-71 | 1980-81 | 1990-91 |
|---|---|---|---|
| Marginal | 35.7 (50.6)* | 50.1 (56.4) | 62.1 (59.0) |
| Small | 13.4 (19.0) | 16.1 (18.1) | 20.0 (19.0) |
| Semi-medium | 10.7 (15.2) | 12.5 (14.0) | 13.9 (13.2) |
| Medium | 7.9 (11.2) | 8.1 (9.1) | 7.6 (7.3) |
| Large | 2.8 (3.9) | 2.2 (2.4) | 1.7 (1.6) |
| All groups | 70.5 (100) | 89.0 (100) | 105.3 (100) |

*Note:*

*Figures in parenthesis represent the percentage of holdings.

Through the overall contribution of fisheries sub-sector is small, development of prolific and fast growing forms of several common fish species coupled with breakthrough in breeding under captivity, fish seed production and multi-layer fish culture has resulted in registering a very high annual growth of 11 percent in aquaculture production during the past decade.

*Table. Area operated - % of total area*

| *Category of Farmers* | *1970-71* | *1980-81* | *1990-91* |
|---|---|---|---|
| Marginal | 9.0 | 12.1 | 14.9 |
| Small | 11.9 | 14.2 | 17.3 |
| Semi-medium | 18.5 | 21.1 | 23.1 |
| Medium | 29.7 | 29.6 | 27.2 |
| Large | 30.9 | 23.0 | 17.4 |
| **All groups (million ha)** | **162.1 (100)** | **163.8 (100)** | **165.6 (100)** |

## AGRICULTURAL BIOTECHNOLOGY INDUSTRY

The top ten pesticide and seed firms for 1997 and 1999. Longer-term time series data on the structure of the chemical and seed company are difficult to obtain. In one of the few published papers to provide this data, Ollinger and Fernandez-Cornejo (1998) provide Concentration Ratio data (CR4) for the United States pesticide industry over the period 1972-1989. During this period, the CR4 oscillates, moving from a high of 50% in 1973 to a low of 37% in 1982, and then rising to 48% in 1989. Although detailed data are not available for the 1990s, there is evidence that the structure of the seed and chemical industry remains dynamic.

The consolidation activity of the ten most active biotechnology firms. Mergers and acquisitions are a key component of the new business relationships that have emerged. As Kalaitzandonakes and Hayenga (2000) note, firm entry in the crop biotechnology industry peaked in the early 1980s, with production innovation continuing throughout the 1980s. Oehmke et al. (1999) also discuss the cyclical pattern in mergers in the biotechnology industry. Aggregate market figures mask the much higher concentration that exists in specific markets. For example, in 1998, Monsanto and Pioneer-HiBred (now owned wholly by DuPont) controlled 15% and 39% of the US seed corn market, respectively. These two same companies controlled approximately 24% and 17%, respectively, of the purchased soybean seed market. For the US cotton market, two companies, Delta & Pine Land and Stoneville, had 71% and 16%, respectively, of the seed market.

The determination of the relevant market concentration is not always done on the basis of output markets. The Federal Trade Commission has used innovation competition to assess the impact of mergers. Examining competition in innovation is designed to focus attention on the impact of a merger on innovative activity. Using data on field trials underway each year, Brennan Pray, and Courtmanche (2000) calculate a CR4 ratio at the R&D stage. In 1988, the four largest firms had 87% of the field trials. The CR4 ratio declined to a low of 63% in 1995, then rose steadily over the next few years to reach 79% in 1998. This concentration at the R&D stage is matched by concentration in terms of the number of patents held. The top four firms held 41% of the corn patents (up to 1996), 53% of the soybean patents (up to 1997), 77% of the tomato patents (up to June 1997) and 38% of the Bt patents (up to 1998).

Concentrated markets do not necessarily imply the presence of market power. Baumol, Panzar and Willig (1992) stress that firms will not be able to exercise market power if the market is contestable. The key requirements for market contestability are: (a) Potential entrants must not be at a cost disadvantage to existing firms, and (b) entry and exit must be costless. For entry and exit to be costless, there must be no sunk costs. Sunk costs are expenditures that cannot be recouped once they are incurred—examples include expenditures made to obtain regulatory approval, expenditures on advertising, and expenditures on R&D.

If there are no sunk costs, potential entrants can use a hit-and-run strategy in which they enter an industry, undercut the price of the incumbents, reap the profits and exit before the incumbents have time to retaliate. In anticipation of entrants acting in this manner, the incumbents forestall entry by keeping price at average cost. The consequence is that, even in an industry that is highly concentrated, prices can be kept at or near competitive levels.

However, if sunk costs are present, firms entering an industry are unable to exit again without losing a portion of their investment. As a result, hit-and-run strategies are much less profitable and incumbents are able to keep price above average cost. Thus, with sunk costs, markets are not contestable and market power is once again an issue.

## Vertical Structure

In addition to becoming larger through growth and horizontal integration, the major agrochemical companies have restructured themselves in other ways.

One of the major changes is increased vertical integration—for example, the combining of seed, chemical and biotechnology activities within the same firm. Although some companies initially tried to include pharmaceutical and biotechnology components within the life science model, practically all of them have now divested their agricultural operations. A further element of this internal restructuring is the increased use of strategic alliances.

Many countries have had their own local seed companies that have over the years developed seed for a specific geographical market and operated sales and distribution systems. The major biotechnology companies are increasingly purchasing these seed companies as a source of seed material in which to insert the genes for herbicide or insect resistance. As an example, in 1997, Monsanto acquired a 30% share of the Brazilian corn seed market with the acquisition of Sementes Agroceres. With its 1998 purchase of Cargill's international seed division, Monsanto now controls over half the Argentine maize seed market.

In 1998, Dow AgroSciences acquired Morgan Seeds, Argentina's second largest corn seed company, and Brazil's Dinamilho Carol Productos Agricolas, another key South American corn seed company. Phytogen (majority owned by Dow Agrosciences) acquired a major cottonseed breeding programme in the Chaco province of Argentina. In 1998, Mexico-based Empresas La Moderna (ELM) bought two South Korean vegetable seed companies and Nath Sluis (agricultural biotech company) of India (Action Group on Erosion, Technology and Concentration [ETC], 2000).

## Factors That Affect Horizontal Industry Structure

Numerous factors are behind the wave of horizontal mergers and acquisitions. Some of these factors are common to all industries and have no specific link to the seed, chemical and agricultural biotechnology industries. Thus, the mergers and acquisitions in the seed and chemical industries are at least in part a result of the need to consolidate costs and rationalize industry capacity, a desire by the management of the firms involved to extend their sphere of influence, and a wish by some firms to pre-empt other firms from taking over valuable assets. MacDonald (2001) provides an overview of concentration levels in a number of US agricultural sectors, many of which have seen higher concentration in recent years.

There are at least two explanations for increasing concentration, however, that pertain specifically to the seed, chemical and agricultural biotechnology industries: (a) sunk costs, and (b) escalation strategies.

## Sunk Costs

Two sources of sunk costs appear to be important in determining industry concentration: (a) intellectual property rights and R&D expenditures, and (b) the regulatory requirements that governments have introduced before the products of the R&D activity can be marketed.

On the empirical side, Ollinger and Fernandez-Cornejo (1998) examine sunk costs and regulation in the US pesticide industry. Using data over the 1972-89 period, they find that research costs and pesticide regulation costs negatively affect the number of companies in the industry, and that smaller firms are affected more strongly by these costs than are larger firms. Research and regulation costs also encourage foreign-based firms to expand into the US market and to force less profitable innovative firms to exit the market. Ollinger and Fernandez-Cornejo also point out that their results on the impact of regulatory costs generally match those found in other industries.

These empirical results can be linked to theoretical arguments concerning sunk costs and the economies of scale and scope that sunk costs tend to create. As Sutton (1991) outlines, the presence of sunk costs means that for firms to be profitable, price needs to be raised above marginal cost, typically by reducing the amount of competition (*i.e.*, the number of firms). Sutton identifies two types of sunk costs: exogenous and endogenous.

Exogenous sunk costs are those beyond the control of the firms in the industry. Regulatory costs are a good example of exogenous sunk costs. In contrast, endogenous sunk costs are firm-level strategic variables. Advertising and R&D are examples of endogenous sunk costs. The rest of this section examines the connections between intellectual property rights (IPRs), R&D expenditures, regulatory costs, and economies of scale and scope.

Economies of scale and scope are major factors in propelling industries towards concentration. Economies of scale exist when average costs fall as more output is produced; economies of scope exist when the total cost of producing two outputs together is less than the cost of producing the two outputs separately.

Because economies of scale and scope mean that larger and diversified firms have lower average costs, there is clearly an incentive for firms to get large. Indeed, those that do not get large are vulnerable to being driven out of the market by larger and more cost efficient firms. Of course, there is a limit to how large firms can get. Although development and production costs may fall with an increase in the size of the firm, other costs—particularly those associated with administration—rise. Nevertheless, economies of scale and scope clearly create pressures for consolidation.

Economies of scale and scope are created as a result of investment in non-rival goods. Unlike rival goods—such as materials, labour, and energy—which can only be used in one place, by one person and at one time, non-rival goods can be used in more than one place and by more than one person, all at the same time. This feature of non-rival goods—that they can be used over and over again—means that output can be increased without having to increase all inputs. Consequently, economies of scale and scope are created.

Much of the expenditure on non-rival goods—examples include R&D and regulation expenditures—can be considered a sunk cost.

Intellectual property, of course, is a good example of a non-rival good. Indeed, ideas generally are considered non-rival goods. It is widely believed that many of the high-technology industries, including the biotechnology and information industries, are subject to increasing returns. Romer (1990) postulates that this is a result of the distinction between physical goods and ideas. Ideas are not scarce, and as such, any industry based primarily on the trade of intellectual property will not face diminishing returns in their primary resource, the idea. The protection of intellectual property, through IPRs, is designed to encourage innovation and to provide incentives for the development and diffusion of new products and technologies.

An example illustrates the connection between intellectual property and economies of scale and scope. Suppose a biotechnology firm has some intellectual property, such as a particular gene that has been isolated. This intellectual property can be used over and over again as the firm expands its activities. For instance, if the company wishes to develop seeds for a new crop, it will not have to invest again in the research that isolated the gene. Although the development of a new seed will require additional lab and greenhouse space, labour, and materials, the expenditure on the technological advancements does not have to be made again.

A similar result occurs if the firm needs to invest substantial amounts of money in obtaining regulatory approval for a seed—although the production of additional units of the seed will require additional costs, the regulatory expenditures do not have to be made again. Once again, large companies typically have an advantage, because they are able to spread the costs of obtaining regulatory approval over more output. Thus, the greater the regulatory requirements in an industry, the more concentrated the industry is expected to be.

Intellectual property may also create economies of scope. Once a specific gene has been isolated—for instance, a gene that confers a resistance to a particular herbicide—this gene can then be put in a number of crops. Once again, the production costs of a number of products together will be less than if the products were produced separately.

To recap, R&D expenditures and regulatory costs are both sunk costs and a source of economies of scale and scope. Because economies of scale or scope mean that larger and more diversified firms have lower average costs, there is clearly an incentive for firms to get large. Indeed, those that do not get large are vulnerable to being driven out of the market by larger and more cost efficient firms.

These theoretical results are supported by empirical observations in the US pesticide industry.

## Escalation Strategies

An escalation strategy is one in which a company spends large amounts on R&D and engages in mergers and acquisitions to achieve a dominant role in the market—that is, the firm tries to leapfrog its competitors to become the dominant firm. Escalation can be a profitable strategy when there is a high degree of substitutability with competitors' products on the demand side, and there are scope economies on the supply side. Both of these factors are present in the agricultural biotechnology industry.

The isolation of a gene that provides particular advantages and which can be inserted into a number of crops means there are scope economies on the supply side.

There are also clear supply side scope economies associated with the enabling technologies that are required to use these genes. And on the demand side, herbicide and insect resistant seeds and the accompanying chemicals are clearly a substitute product for traditional seeds and herbicides and pesticides. Firms engage in escalation strategies in a number of ways.

Although internal growth through R&D is one way, the more common method is through consolidation via horizontal mergers and acquisitions. As the theory predicts, the combination of demand substitutability and supply side scope economies present in the seed and chemical industry appears to be linked with escalation strategies.

Monsanto has clearly followed an escalation strategy, with Dow, Novartis, and others following similar strategies of making acquisitions and spending significant amounts on R&D.

## CROP PRODUCTIVITY

Global demand and consumption of agricultural crops for food, feed, and fuel is increasing at a rapid pace. This demand for plant materials has been expanding for many years. However, recent increases in meat consumption in emerging economies together with accelerating use of grain for biofuel production in developed countries have placed new pressures on global grain supplies. To satisfy the growing, worldwide demand for grain, two broad options are available:

(1) The area under production can be increased or

(2) Productivity can be improved on existing farmland.

These two options are not mutually exclusive and both will be employed to produce the additional 200 million tonnes/year of corn (Zea mays) and wheat (Triticum aestivum) estimated to be needed by 2017. Both options will alter the environmental footprint of farming. Of the two options, increasing productivity on existing agricultural land is preferable as it avoids greenhouse gas emissions and the large-scale disruption of existing ecosystems associated with bringing new land into production.

In the United States, breeders, agronomists, and farmers have a documented history of increasing yield. U.S. average corn yields have increased from approximately 1.6 tonnes/ha in the first third of the 20th century to today's approximately 9.5 tonnes/ha. This dramatic yield improvement is due to the development and widespread use of new farming technologies such as hybrid corn, synthetic fertilizers, and farm machinery. The introduction of biotechnology traits and development of new breeding methodology using DNA-based markers are further improving yields. Outside the United States, similar farming practices have been adopted in some agricultural nations, but in many major grain-producing countries, yields still lag well behind world averages. By continuing to develop new farming technologies and deploying of them on a global basis, demand for feed, fuel, and food can be met without the commitment of large land areas to new production.

## CROP EVOLUTION

Alphonse de Candolle, the founder of the study of crop evolution, was primarily interested in biogeography, the study of distribution of plants in relation to their environment. In his opinion, among the factors influencing this distribution are 'historic' factors such as glaciation events that are not related to the intrinsic adaptation characteristics of the species. When he considered crop plants, he posited that one particular historic event - domestication - had had a paramount importance in determining their adaptation and, hence, distribution, as well as their genetic characteristics Four types of evidence could be utilized to determine the origin of domestication of crop plants These were archaeological (or more specifically archaeobotany), botanical (*i.e.* the distribution of the wild, ancestral relative), historical (or the existence of a written record documenting the existence or importance of the crop) and linguistic evidence (*i.e.* the existence of words designating the crop or objects or concepts related to the crop in native languages).

Of these four types, the first two provide the most abundant and reliable evidence. Since the time of de Candolle, however, additional scientific methods have increased considerably our power to determine centers of crop domestication based on archaeological and botanical arguments

Although common bean is not a species originating in the Near East, the focus of the present symposium, it presents nevertheless an interesting domestication history, which is only recently being elucidated. The first botanically accurate description of common bean in European herbalsWhereas barely 50 years had passed after the conquest of the Americas, the actual origin of the crop had already been lost to the writer of the herbal and later European botanists. Linnaeus assigned the origin of common bean to India; de Candolle himself expressed doubts that common bean had been domesticated in the Americas. It was only during and after the Second World War that it became generally admitted that common bean originated in the Americas. The conclusion was based on the discovery of wild common bean in Argentina and Guatemala and archaeological remains in the Americas

Based on the extensive distribution of the wild relative speculated that common bean originated through multiple domestications as did other crops originating in the Americas, such as cotton, pepper and amaranth. observed that there was a Mesoamerican cultivar group with smaller seeds and an Andean group with larger seeds. None of these observations, however, provided substantial experimental evidence and hence a clearer picture of the actual process of domestication for this species. For example, one could ask whether the two cultivated groups identified resulted from distinct domestications or a single domestication event followed by divergence into a small-seeded and a large-seeded group. One could also ask where these single or multiple domestications took place. The archaeological record of the origins of agriculture in the Americas is rather limited. In Mesoamerica, This important type of information therefore pales in comparison with what is available in the Near East In part, this may be due to the relatively humid environment required by many crop relatives in the Americas, which render long-term conservation of crop remains less likely. Therefore, in relative terms compared with archaeological information, there is a relatively higher emphasis on botanical evidence to decipher crop evolution in the Americas than in the Near East.

The use of molecular marker information has increased our understanding of the domestication process in common bean a crop originating in the Americas. In addition, we hope to demonstrate how molecular analyses can bridge the gap that has separated botanical and archaeological evidence, the twin pillars on which rests the study of crop evolution. The power of molecular biology to change scientific paradigms is by no means unique to this field but extends to many other fields as well.

## General Methodological Principles

Our current understanding of the origins of domestication in common bean is based on a two-pronged approach relying on the study of wild-growing *P. vulgaris* beans, on one hand, and the use of biochemical or molecular markers, on the other. The use of wild beans is predicated on the fact that their dispersal occurs over much smaller distances (at most a few meters after each vegetation cycle, if at all) than their domesticated descendants. The latter are subject to long-distance exchange, trade, gift or human migration, all of which tend to obscure the original geographic distribution patterns of genetic diversity. Recent explorations have provided a more complete representation of the actual distribution of wild *P.* beans are distributed from approximately 30°N Lat. to 30°S Lat. at mid-elevations from 1000-2500 m asl in regions with moderate rainfall. A common characteristic of the environments in which wild beans are found is the existence of a marked dry season, an important feature needed for seed dispersal.

Genetic relatedness is traditionally assessed with morphological phenotypic traits. Plants with a similar phenotype are inferred to be genetically related either through common ancestry or through geneflow (hybridization). Occasionally, however, inferences based on phenotypic similarity are erroneous. Plants can arise from different selection episodes leading to the same phenotype (convergent evolution). In addition, similar phenotypes can have a different genetic basis (genetic heterogeneity). In both cases, an inference of genetic relatedness would be unwarranted. In contrast, biochemical, and even more so, molecular markers provide information on genetic relatedness that is generally (although not always) devoid of selective influences.

All biochemical and molecular markers were created equal! Differences have been observed with regard to the ease of analysis, reproducibility, level of polymorphism, number and genome distribution of loci Of particular note is the molecular basis of polymorphisms. The more complex the basis, the more desirable the marker, because complexity will reduce the probability of repeat mutations, especially in conspecific materials, which are, by definition, closely related. For example, polymorphisms based on DNA nucleotide substitutions are simple changes that could presumably occur repeatedly. Analyses based on single or few RAPD or RFLP markers are therefore unreliable in determining genetic relatedness that sample a large number of changes at the molecular level, because they include either a large number of markers or complex markers such as actual DNA sequences or multigene families, are more robust with regard to repeated changes. For example, banding patterns of seed proteins such as phaseolin in common bean, when analyzed in denaturing SDS-polyacrylamide gel electrophoresis, represent a compomy image resulting from changes at the DNA nucleotide level (duplication of genes, within-gene duplications, nucleotide substitutions) and at the translation level (co - and post-translational modifications). Therefore each pattern is unlikely to have arisen more than once. As a consequence, genotypes that exhibit the same protein pattern are likely to share a common ancestor

One of the major experimental difficulties in crop evolutionary studies is the distinction one has to make between ancestry (domestication) and geneflow as a cause of genetic relatedness. This is particularly the case given that wild ancestors and domesticated descendants belong to the same biological species and can therefore freely hybridize. Hence, similarity between ancestors and descendants even at the molecular level cannot be assumed to be due to domestication For example, DNA sequence data and the utilization of an outgroup (*e.g.* closely related species) can shed additional light on this problem because they can provide information on the nature of the polymorphism and the direction of the change, *i.e.* the ancestral vs. descendant states of the molecular polymorphism.

## *Domestication Pattern*

The origin of the *Phaseolus vulgaris* progenitor in its wild state can be traced to the Pacific slope of Ecuador and northern Peru, based on the distribution pattern of direct repeats in phaseolin seed protein genes Phaseolin is coded by a small multigene family inherited as a single Mendelian unit and located on linkage group B7 The phaseolin locus is at most 190 Kb long and contains 6 to 9 genes Differences among these phaseolin sequences include nucleotide substitutions and the presence or absence of tandem direct repeats. Three repeats have been identified so far: a 21 bp repeat in the third intron, a 15 bp repeat in the fourth exon and a 27 bp repeat in the sixth exon. The existence of these tandem repeats provides an interesting landmark to follow the evolution of phaseolin genes. Indeed, the generation of a repeat is a more likely event than the precise loss of such a repeat, which would be needed to restore a sequence without repeats. Hence, phaseolin genes without repeats represent probably a more ancestral state than phaseolin genes with repeats. This was verified by analyzing the presence of tandem repeats with a polymerase chain reaction (PCR) test of phaseolin sequences in two subspecies of a related species, *P. coccineus* subsp. *polyanthus* and *P. coccineus* subsp. *coccineus*. In these two taxa, none of the three repeats could be identified, suggesting that the repeatless state is indeed the ancestral state. A survey was then conducted among wild *P. vulgaris* to determine if any phaseolin haplotypes lacked genes with tandem repeats. Only wild bean populations from Ecuador and northern Peru on the Pacific slope of the Andes appeared to be devoid of phaseolin genes with tandem repeats It was inferred that these wild populations therefore represent the presumed ancestor of the species. This conclusion is somewhat unexpected as most wild *Phaseolus* species are distributed in Mexico. One could have thought that *P. vulgaris* would have originated in that region as well and that other wild *P. vulgaris* populations distributed elsewhere were ultimately derived from Mexican ancestors. It may be, however, that some wild *Phaseolus* ancestors ('proto-vulgaris' and other species) were dispersed from Mexico to Central America and the Andes to give rise to locally distributed wild *Phaseolus* species, such as *P. costaricensis* From such a local distribution, a species such as *P. vulgaris* could have acquired a secondary, broader distribution by dispersal towards the north and south. It is worth remembering that lima bean (*P. lunatus*) also shows an extensive distribution of its wild relative from Mexico to Argentina and that the species consists of two genepools, an Andean genepool originating on the Pacific side of the Andes in Ecuador and northern Peru (although at lower altitudes than wild *P. vulgaris*) and a 'Mesoamerican' genepool distributed from Mexico to the southern Andes. Although the domestication region of the 'Mesoamerican' genepool is unknown so far, it is likely to be located somewhere in Mesoamerica, given that most Mesoamerican cultivars are located there

From the small nuclear area in Ecuador and northern Peru, wild *P. vulgaris* were dispersed towards both the north (Colombia and Venezuela, Central America and Mexico) and the south (southern Peru, Bolivia and Argentina) to achieve their current distribution In a subsequent step, multiple domestications took place in the Mesoamerican and Andean regions. At least one domestication may have taken place in west-central Mexico (Jalisco), a conclusion based on phaseolin data. Most Mesoamerican cultivars exhibit the same phaseolin type ('S'-type). Wild beans with the same phaseolin type and without morphological signs of past hybridizations to domesticated beans are concentrated in the Mexican state of Jalisco, Interestingly, Doebley *et al.* suggested a similar domestication area for maize.

A second major domestication took place in the southern Andes although a more specific region awaits confirmatory. Phaseolin suggest possible multiple domestications within the Andean region because at least four phaseolin types have been identified among Andean beans. Alternatively, the original populations that were domesticated were polymorphic for phaseolin. Further are needed to distinguish between these possibilities. A third, likely minor domestication may have taken place in Colombia, although regions in Central America deserve further study because samples of wild *P. vulgaris* from that area were unavailable.

## DIVERSIFICATION AND CROP INCOME STABILITY

The effectiveness of diversification as a self-insurance strategy. Stability is measured in a relative context and is identified with a low coefficient of variation for household crop income. Crop income is measured in two ways, gross returns and net returns to family owned resources which include land, capital, family labour and management. All non-marketed outputs and inputs are valued at their respective opportunity costs in measuring gross and net returns.

Of the 180 cultivator households in the VLS sample, 134 have data on crop income for each of the five years from 1975-76 to 1979-80. The majority of the other 46 households leased out their land for one or more cropping years; a minority left the sample and were replaced.

Therefore, there are 134 cultivators for whom coefficients of variation of household gross and net crop returns are estimated from 1975-76 to 1979-80. CVs are calculated for each household across the five cropping years. It was also expected that crop income stability is a function of some of the resource base determinants. Moreover, differences in crop income stability among farmers can also be attributed to fluctuations in the resource base across cropping years. To assess the impact of diversification on crop income stability, it is important to control for shifts in cultivated area, irrigated area, and land quality over time. This is particularly true for regions such as Sholapur where tenancy is prevalent (Jodha, 1981).

Descriptive information on the CVs' of the dependent and independent variables. The size of the CVs' at about 40 to 50% is another indication of the riskiness of dryland agriculture in the semi-arid tropics of peninsular India. Fluctuations in gross returns are somewhat less than in net returns. Somewhat surprisingly, both average gross and net returns are less stable in the more heavily irrigated Mahbubnagar region compared to the dryland Sholapur and Akola regions.

It is believed that crop income stability can be traced not only to the level of the resource base but also to its intertemporal variation. Because crop and plot diversification are themselves functions of the same resource base viewed in the context of association rather than causation.

It is perhaps heroic to try to explain inter-farm variation in crop income stability within a narrow ecological region based on only five year data. Nonetheless, some of the hypothesized determinants behave as expected and have statistically significant estimated coefficients. Crop diversification is negatively and strongly associated with crop income instability in the more dryland Sholapur and Akola regions. A 10 percent proportional increase in the crop diversification index is associated with 4.8 and 13.0 decrease in the CV of net crop returns in Sholapur and Akola respectively. At the margin, crop diversification is about three times more effective in stabilizing net returns in rainfall assured Akola than in drought-prone Sholapur.

In Mahbubnagar, plot diversification contributes positively to crop income stability. In contrast, in the Akola and Sholapur villages, the results mildly suggest that plot diversification is associated with greater crop income instability. Production conditions for a crop in these more dryland villages must be quite narrowly defined as diversification across space with the same commodity does not dampen fluctuations in income. This is particularly true in Sholapur where field diversification is positively correlated with gross and net return instability. These results support Walker and Singh who found land sharecropped in Sholapur was of inferior quality and was more likely to be the site of crop failure than owner-operated land. Plot diversification appears to be a resource management strategy rather than an action that offers scope for crop income protection.

As expected, the CVs' of the resource base determinants are for the most part-the exception being land value in Mahbubnagar-positively associated with crop income instability. Larger farms having more gross cropped area are characterized by greater crop income stability in the three regions. However, the impact of farm size on crop income stability is not large.

Evaluated at their arithmetic means, a 10 percent increase in gross cropped area leads on are average to about a 2 percent reduction to the CV of net returns.

The evidence with respect to irrigation is not as clear, but it weakly suggests that greater access to irrigation is accompanied by greater instability in net returns particularly in the Akola villages where irrigated land accounts for only 3 percent of the gross cropped area. Besides following inter-regional differences were also observed:

1. The impact of irrigation on crop diversification and income stability hinges upon cropping conditions and the existing level of irrigation.

When an expansion in irrigation takes place for rainy season paddy production in a region where many farmers have some access to irrigation, irrigation leads to more specialization in paddy production. In contrast, when an investment in well irrigation occurs in a multicrop predominantly dryland region, it enhances opportunities for further crop diversification particularly in the postrainy season. More profitable production opportunities in turn translate into greater relative variance in net crop returns income and increased crop income instability. The net effect of irrigation on crop diversification and income stability depends on the source of irrigation, ecological location, and level of growth of irrigation facilities.

2. Crop and plot diversification fluctuate more in more drought prone regions like Sholapur. Crop diversification appears to be more sensitive to agro-climatic determinants than plot diversification.
3. Different regional ecological conditions offer differential opportunities for risk averse farmers to diversify. Therefore, it is not surprising to find that risk attitudinal differences vary regionally in their effect on diversification. Where they are significant, differences in risk attitudes do not exert a sizeable influence on decisions on diversification; their proportional impact is about the same as differences in draft power availability.
4. Crop diversification effectively imparts stability to net crop returns in the dryland farming Sholapur and Akola villages. In the more highly irrigated Mahbubnagar villages, income instability and crop diversification are not significantly associated.

The implications of these conclusions are threefold. First, at present levels of technology it appears that increasing land fragmentation associated with plot diversification does not represent a significant cost in terms of income foregone or a material benefit in the form of enhanced crop income stability.

Thus a prospective land consolidation programme would not compromise the risk-bearing capacity of farmers, nor would it significantly increase average crop income of farm households at existing levels of technology. These issues require closer scrutiny within a more rigorous household economics framework.

Secondly, the emergence of crop diversification as a strong force in protecting net crop returns in the more rainfall-assured Akola region decreases the demand for government risk-reducing policies, such as crop insurance, because farmers can rely on self-insurance through diversification to enhance crop income stability. These results further underscore the emerging consensus that the potential benefits of crop insurance programmes will be greater in more risk-prone, dryland regions like Sholapur for three reasons. First, yield variability is the predominant source of risk in these regions (Barah and Binswanger, 1982), yields are more likely to be positively covariate across farmers which augments the efficiency of a homogeneous area approach to dampen fluctuations in crop income and third, there is less scope for crop diversification to be an effective self-insurance measure.

Lastly, an investment in irrigation may have a negative effect on relative crop income stability when instability is measured with respect to the coefficient of variation. The farm-level effects of irrigation may be markedly different from what obtains at the district, state or national levels.

## CROP ROTATIONS FOR INCREASED PRODUCTIVITY

Crop rotation is a planned order of specific crops planted on the same field. Crop rotation also means that succeeding crops are of a different genus, species, subspecies, or variety than the previous crop. Examples would be barley after wheat, row crops after small grains, grain crops after legumes, etc. The planned rotation sequence may be for a two-or three-year or longer period. Some of the general purposes of rotations are to improve or maintain soil fertility, reduce erosion, reduce the build-up of pests, spread the workload, reduce risk of weather damage, reduce reliance on agricultural chemicals, and increase net profits.

Crop rotations have fallen somewhat into disfavour because they require additional planning and management skills, increasing the complexity of farming. A shift away from livestock programmes in most parts of North Dakota has also reduced the need for pasture and hay crops and eliminated some rotational crops such as alfalfa from many farms. Solid-seeded crops such as small grains and flax have predominated in the past, but row crops such as sunflower, pinto bean, corn and soybean provide additional planting options and more reasons for crop rotations.

### General Effects of Rotations

One immediate economic benefit of crop rotations is improved yields. For example, sunflower yields over eight years at Crookston, Minnesota (24) were often significantly greater in rotation with other crops than when continuous sunflower was grown. Wheat yields were also greater with rotation than continuous wheat in an eight-year study (22) conducted with different crops at Fargo. A study at the Agriculture Research Service at Mandan has shown that increased hard red spring wheat yields can be expected when an alternative crop is included in the rotation (27).

Rotating to a different crop such as wheat on barley ground usually results in higher grain yields when compared to continuous cropping of wheat. Even greater benefits are usually obtained by rotating two distinctly unrelated crops, such as a small grain seeded into land where the previous crop was a legume or other herbaceous dicot such as flax or sunflower. Many of the reasons for the beneficial effects of rotations are not completely understood. Some of the more important beneficial effects that can be obtained from a well planned crop rotation are:

- Reduced insect and disease problems
- Beneficial residual herbicide carryover
- Improved soil fertility
- Improvements in soil tilth and aggregate stability
- Soil water management
- Reduction of soil erosion
- Reduction of allelopathic or phytotoxic effects.

Pest control is often an important reason for crop rotation. Rotations can be used to prevent or partially control several pests and reduce the reliance on chemical and mechanical control. A combination of crop rotation and pesticides is often more effective in reducing pest populations to economic levels than pesticides alone. Pesticides that provide economical control are not available for pests such as white mold in potato, dry bean, and sunflower, and crop rotations are the only feasible control method for reducing the impact of sclerotinia.

### *Disease Control*

Crop rotation has tremendous potential for reducing and often preventing the transmission of disease. Disease pressures change with changing environmental conditions. Crop rotation, in combination with cultural practices plus necessary fungicides, is the most desirable method of disease control.

### *Soil Nitrogen*

Legumes in the rotation can be used to increase the available soil nitrogen. Symbiotic nitrogen-fixing bacteria called rhizobia form nodules on the roots of legume plants and convert or fix atmospheric nitrogen to organic nitrogen. The amount of nitrogen fixed varies with species, available soil nitrogen, and many other factors. Fixed nitrogen not removed from the land by harvest becomes available to succeeding crops as the legume tissues undergo microbial decomposition. When the legume crop is seeded, rhizobia inoculum should always be applied to the seed to ensure the most productive commercial strains are available to form nodules and that inoculating bacteria are always present. Even though indigenous bacteria may be present in the soil, research shows improved commercial strains of rhizobia have more capacity to fix nitrogen. Legumes have the capacity to fix large amounts of nitrogen. Research in Minnesota (16) indicated that alfalfa fixed an average of 172 pounds of nitrogen annually during the first two years of production. However, only a portion of the nitrogen fixed is available to the next crop because much is removed in the harvested alfalfa.

### *Insect Control*

Insects which can be controlled entirely or in part by rotations. In addition, insect populations may become greater within a region where only one or two crops are continuously grown in contrast to a region where several crops are grown in rotation. Insects such as corn borer, sunflower seed and stem weevil, and many others readily migrate to nearby or distant fields. Therefore, only partial control can be obtained by rotation. Increasing field isolation from fields seeded to the same crop the previous year will often increase the effectiveness of crop rotation as an insect control method.

### *Weed Control*

Rotations can be used to cause shifts in weed populations. Populations of certain weed species can be suppressed by competition from the crop raised or by the selective use of herbicides. Wild mustard populations can be reduced by selective treatment of small grain grown in rotation with row crops. Grass weed populations, often a problem in small grains, can be reduced by the use of the appropriate herbicide in the previous row crop.

Herbicides can have both beneficial and harmful residual effects on the next crop. Therefore, planning the correct sequence of herbicide use together with crop selection has become a necessary part of rotation management.

### *Soil Tilth and Structure*

Many farmers who rotate crops comment on the improvement in tilth or friability of soil following soybean or other row crops. In a Colorado study (25) involving corn, sugarbeet, and barley planted on succeeding years, soil aggregate stability was increased from 67 to 76 percent when three years of alfalfa were added to the rotation.

Increased aggregate stability reduces the tendency of the soil to puddle or crust, increases rate of water infiltration under certain conditions, and may also reduce wind erosion.

### *Soil Moisture*

Crop rotation can lead to greater overall efficiency in soil water utilization. Spring seeded small grains usually deplete soil water 3 to 4 feet deep. In contrast, sunflower, safflower, corn and sugarbeet are deep-rooted crops which can deplete soil water to depths of 5 to 6 feet. Therefore, deep-rooted crops such as sunflower following small grains can take advantage of the extra reserve of deep moisture and also any nitrogen which was positionally unavailable to a shallow-rooted crop.

Alfalfa and sweetclover, also deep-rooted crops, can be used to dry up saline seeps and other wet areas. Depletion of soil water in saline areas prevents the accumulation of salts on the surface, permits movement of the salts downward by leaching, and allows recropping to a cash crop such as wheat. Alfalfa or sweetclover should be seeded on the upslope recharge area to use and remove soil water as deeply as possible from the soil profile, reducing lateral flow of water and salts into the saline seep discharge area.

### *Reduction of Soil Erosion*

Crop rotations combined with recommended tillage practices can play an important role in reducing wind and water erosion. Solid seeded crops such as small grains provide more protection against water erosion than row crops. Permanent crops such as hay or pasture provide even more protection against erosion. Management of crops to provide sufficient residue throughout the year is essential for satisfactory control of both wind and water erosion. No-till or minimum till farming is highly desirable as a conservation practice, but crop rotation must be used to reduce the buildup of insect, disease and weed pests. Common pest problems associated with continuous no-till wheat have been *Fusarium* head blight (scab), wheat streak mosaic, root rots, tan spot, wheat stem sawfly, increases in winter annual grass species, and serious infestation of perennial noxious weeds such as quackgrass and Canada thistle. Other crops grown continuously have had similar types of problems.

The effect of various crop rotations (29) on soil erosion is vividly. The addition of barley and hay or pasture to the rotation was estimated to reduced the degree of soil erosion by more than 50 percent when compared to continuous corn. The expected reduction in soil erosion would be even greater on steeper slopes.

### *Allelopathy-phytotoxicity*

The reasons for improved yields due to crop rotations are not completely understood. Research has attempted to reveal some of the unknown factors.

Terms such as phytotoxicity, allelopathy and autotoxicity are coming into common usage. Phytotoxicity, a general term, is defined as chemical that is toxic to plant growth whether it is derived from plant products or synthetic (herbicide or other pesticide residues). Allelopathy refers to plant material or chemicals derived from these materials which inhibit the germination, growth or growth of another species. Autotoxicity refers to plant material which inhibits the germination, growth and growth of the same species.

Legumes such as alfalfa, while often beneficial to non-related species, exhibit autotoxic effects to alfalfa seedlings. Research in Illinois indicated older stands caused greater inhibition of new seedlings. This study (21) indicated that one year out of alfalfa was sufficient to nullify the detrimental effects of the alfalfa.

### *Rotational Benefits to Corn and Sorghum*

Corn yields following alfalfa and soybean, even with 300 pounds of applied nitrogen per acre, were greater than with continuous corn. The beneficial effects of alfalfa on corn yields during the second and third years following alfalfa. Even with 75 to 150 pounds of added nitrogen, the residual benefits of alfalfa are still evident. First-year corn yields were the same after either soybean or alfalfa.

### *Sugarbeet Rotations*

Yields and quality are usually highest when sugarbeet follows barley or wheat in the crop rotation. Yields are usually high when sugarbeet follows corn, potatoes or summer fallow in rotation, but higher than desirable soil nitrogen levels may reduce crop quality. Three years research in Minnesota indicates sugarbeet yielded significantly less when following soybean versus barley in rotation. One year of research indicates sugarbeet yields were reduced following dry edible beans in rotation. Several possible explanations for reduced yields when sugarbeet follows soybean include: (1) herbicide carryover; (2) lack of available soil water; (3) increased rhizoctonia root rot; or (4) alleopathic effects.

Typical Red River Valley sugarbeet rotations include: Sugarbeet — soybean or dry edible bean — small grain-small grain or, if using a three year rotation, only one year of small grain. In areas where corn is more common, a desirable rotation is sugarbeet — soybean — corn. Soybean following sugarbeet is desirable because the soybean may adjust to periods of moisture stress better than small grains following sugarbeet. Planting small grains prior to sugarbeet reduces root rot potential in the sugarbeet. Early harvested small grains also allow time for sugarbeet growers to soil sample, apply fertilizer, and prepare seedbeds for the subsequent sugarbeet crop.

### *Sugarbeet Rotation Guidelines*

- Three year rotations are the absolute minimum length of an acceptable rotation.
- Four year or longer rotations are desirable to minimize root disease, Cercospora leaf spot and herbicide carryover.
- Avoid canola in crop rotations with sugarbeet if at all possible, because it is an alternate host for sugarbeet cyst nematode.
- Avoid herbicide carryover that may damage sugarbeet or rotational crops.
- Manage fertilizer nitrogen use throughout the rotation.
- American Crystal Sugar Company data shows a trend towards more sugar production per acre in rotations of five years or more.

## MONSOON CRISIS AND CROP PRODUCTION ESTIMATES

While it is a fact that late monsoons, drought, and floods had an impact on food production in the country, its influence on prices for commodities like rice had been nominal (as compared to other commodities). This is primarily due to excellent food grain production last year that immensely aided in boosting government stocks. In the current crop season, however, crop estimates point to a gloomy picture.

India's overall food grain production in Kharif 2009 is estimated to come down by 16 percent with significant downfalls in Rice and Coarse Cereals respectively. Moreover, in the case of Pulses, estimates are down by 7.5 percent, mainly due to low rainfall in central, western and northern India. Seasonality issues also came to the fore in the case of fruits & vegetables, notably Potato. Large tracts of Potato were destroyed in West Bengal & Bihar (two main potato growing regions) due to late blight infestation. As a result, strains on food reserves in the country inevitably become evident, and hence the inflationary tendencies. Such a scenario, besides stoking inflationary fear, is also instilling a fear that the country would be unable to manage satisfactory growth in agriculture GDP.

An optimistic counterview is being put forth repeatedly which suggests that Rabi harvest would satisfactorily suffice for the shortage of food articles and would also cover-up for the losses sustained so far. Estimates however, do not support this proposition. Experts point out that it is the third quarter, when Kharif harvest takes place and the fourth quarter (for Rabi) that are extremely important for the Agriculture GDP of the country. These two quarters alone account for nearly 60 percent of the agriculture GDP, with Kharif accounting for nearly one-third and Rabi, a little over than a quarter. As Kharif estimates points to a sharp decline, such a scenario puts forth immense pressure on Rabi season to match up. However, this is an onerous task; the country would require substantial growth figures in fourth quarter. To elucidate, Widely consumed pulses varieties such as Moong, Tur are mainly kharif crops and their production has already declined; wheat production already has reached a record level of 81 Million tonnes last year and to make up for the loss this year, wheat production has to reach 93 Million tonnes, which is nearly impossible. Besides, oilseeds will have to grow by 26 percent to maintain overall oilseeds production at last year's level. Consequently, the substitution effect that Rabi harvest can have will be limited; it is highly unlikely that improvement in Rabi food grains would cover-up losses in kharif. Shortages in the availability of many commodities seems imminent this season and food inflation is going to have a longer horizon than expected. Here, it is equally important to present an interesting view, which states that we cannot confidently attribute the price increases beign witnessed currently to commodity shortages.

*Can we fully attribute this prices increase to shortage of food-grains and vegetables?* It is pertinent to mention that we have been having two consequently good agriculture seasons prior to 2009; our stockpiles, especially of food-grains is at more than optimum levels. A closer look at half yearly GDP data yields another interesting observation and brings in an element of doubt on attribution of inflation singly to poor production estimates & figures. An interesting point of view has come up, which states that in real terms agriculture GDP as a percentage of private consumption expenditure at nearly 24 percent, is constant for first half of both 2008-09 and 2009-10. Essentially, this means that agricultural supply as a share of demand in real terms is constant. It is not the price for a single commodity that is rising but all agricultural prices covering food grains and fruits and vegetables. These prices are rising when the share of output to demand in real terms is constant, it is not falling. Accordingly it is premature to state that inflation is stemming from supply constraints. On the other hand, the focus of policy makers is to attribute price rise to kharif grains, but this might be fallacious as the share of grains in both price and in the output indices is pretty low, in fact around a sixteenth. We cannot confidently say that there is supply crisis in food products, and that inflation is a result thereof.

## BIOTECHNOLOGY FOR SOCIETAL DEVELOPMENT IN INDIA

The application of biotechnology for societal development has always received special attention since it benefits socially and economically disadvantaged sections of the society.

Through this programme a number of projects are implemented for economic and technical empowerment of the beneficiaries. Efforts are made through training, demonstration and appropriate R&D inputs for providing tailor made solutions for location specific problems of SC/ST, women and rural population using modern biology and biotechnology. The following pages provide the highlights of the achievements.

### Programmes for SC and ST Population

Biotechnology-based Programme for SC/ST population has benefited around 15,000 people directly and about 45,000 indirectly through 54 projects during the year. The projects have been supported in the areas of cultivation of medicinal and aromatic plants, biofuels, production of biofertilizers and biopesticides, organic farming, aquaculture, mushroom cultivation as well as human healthcare interventions etc.

The universities, public funded institutions, Krishi Vigyan Kendras and voluntary/non-governmental organizations have been involved in the implementation of the projects. The target groups have received the benefit of hands-on training and field demonstrations.

Vermicomposting: MR Morarka GDC Rural Research Foundation Jaipur implemented a project in Jalgaon district of Maharashtra, to benefit 2600 SC/ST target population on economic utilization of waste, agriculture residues and cow dung. Over 1780 MT of waste, agricultural residues and cow dung was utilized for vermicomposting through establishment of 28 community facility centres.

Beneficiaries are utilizing vermicompost on crops like onion, sugarcane, cotton, banana, papaya and wheat. Some farmers switched over from conventional crops to cash crops such as colorful capsicum, gerbera as well as inter cropping of safed musli, turmeric and ginger in banana plantation. Farmers have been trained to use bio-fertilizer, foliar sprays and mulching to use the agro-waste for fertility and management.

Marketing linkages have been established with "Organic Shoppe", through which farmers are getting 10-20% premium for their organic produce. About 20 entrepreneurs have taken up vermiculture as commercial activity. extension Material in the local language was distributed to the farmers.

Mushroom cultivation: A project on promotion & production of essential oil, mushroom and vermicompost was implemented by RRL, Jorhat for revenue generation of the ST population of Arunachal Pradesh. Motivation and demonstrationcum-training programmes have been conducted in 13 villages of five districts on cultivation of citronella, production of mushroom and vermicompost. Demonstration units have been established at RRL Branch Naharlagun farm. Mushroom spawn, earthworms and citronella saplings were distributed to the beneficiaries. Several beneficiaries established vermicompost units on commercial basis and are selling vermicompost. The beneficiaries at Ziro, Jairampur, Doimukh, Manmao, Hetlong and Tengman have established oyster mushroom cultivation units. So far, 10 acres of land have been planted with citronella and 7 acres of land with patchouli at Sonajuli village of Papum Pare district.

Medicinal plants: A project on strengthening health care practices and conservation of medicinal and nutritious plants was continued at Herbal Folklore Research Centre, Tirupati for implementation in 100 villages of two Mandals in Andhra Pradesh. Training workshops were organized with the help of Yanadi tribal people for community health volunteers on identification and sustainable harvesting methods of 20 species of locally used medicinal and aromatic species. The reputed healers were encouraged to prepare medicines needed for PHC from the identified plants. About 6000 seedlings were distributed to the beneficiaries for planting in kitchen gardens. Green health kits (samples of herbal medicines) and educational material in local language containing identification, cultivation techniques and preparation of medicines with their therapeutic properties were distributed to the trainees.

A project on creating awareness and promotion of medicinal plants for the production and marketing was implemented by RRL, Jorhat to benefit ST population of Arunachal Pradesh. Awareness and training programmes were organized on the importance, scientific cultivation and marketing of Piper longum, Kaempferia galanga, Andrographis paniculata, Stevia rebaudiana, Rauvolfia serpentina and Mentha piperita. Three nursery-cumdemonstration centers were established at Roing (1000 ft msl), New Bomdila (8160 ft msl) and Ziro (4350 ft msl) and 19000 saplings were distributed to ST farmers/ NGOs for introduction and propagation in their fields. At high altitude, villagers were advised to cultivate Aconitum heterophyllym, Coptis teeta and Picrorrhiza kurroa.

Beneficiaries have stopped collecting wild Homalonema aromatica, an important but rare aromatic plant of Arunachal Pradesh, and have started its cultivation in their farms. Ten community nurseries have been established with the help of Vivekananda Trust, Roing at ten villages of Lower Dibang Valley District. Twenty farmers were trained on cultivation and post-harvest management of medicinal plants and marketing tie-up has been established. Around 1266 local ST people participated in the awareness programmes.

A project on organic cultivation of medicinal and aromatic plants for income generation and sustainable development in Uttarkashi District of Uttaranchal was undertaken through Forests Conservation Programme of World Wide Fund for Nature, New Delhi. Demonstration-cum-Resource Material Production Centre established for training on vermicomposting and cultivation of species. Construction of poly house, vermi-composting units and planting of seeds of Rauvolfia serpentine completed. Awareness and motivation camps were organized for the farmers. Extension material brought out in the local language for cultivation of Rauvolfia serpentina, Coleus barbatus and Pelargonium graveolens were distributed to the farmers.

A project on developing skills and economically sound vermicompost units utilizing distillation waste of Mentha arvensis (menthol mint) was implemented for the benefit of SC/ST communities of Lucknow and Barabanki Districts of Uttar Pradesh. Around 200 farmers were trained on menthol mint nursery raising, transplanting, distillation and practices followed in vermicomposting. Demonstrations were carried out in farmers' field using high yielding variety 'Kushal'. Farmers trained on organic farming started vermicompost production using distillation and other agro-waste. The yield data obtained from the demonstrations showed the possibility of reduction of chemical fertilizers by 75% without affecting the yield.

## Biofuels

A project was continued on cultivation of Jatropha in arid lands of Namakkal District in Tamil Nadu for production of biofuel and other products for SC/ST community by Centre for Research in Social Sciences, Coimbatore. A high tech nursery was established and plantlets of Jatropha were raised. Training programmes were organized for 223 members on cultivation of Jatropha in 31 acres of land. About 30,000 plants were supplied for cultivation. Awareness camps, training and demonstrations were conducted on Jatropha cultivation, oil extraction and soap making. SHG groups were trained in nursery techniques. Women self help groups were trained in various activities of jatropha cultivation and soap making. Extension material bought out in the local language was distributed to the beneficiaries.

Biocontrol agents: A project was implemented on development of eco-preneurship among the coastal community of Pondicherry by MS Swaminathan

Research Foundation, Chennai. Sixteen SHGs involving women and youths were formed in Ariyankuppam, Nonankuppam, Kakayanthope and Veerampattinam villages. Demonstrations and trainings were conducted to educate trainees on basic aspects of IPM, importance of biocontrol agents, production and marketing of Trichogramma, importance of human nutrition, role of edible mushrooms and production & marketing of oyster mushroom. Micro-enterprise was developed for mushroom and Trichogramma production. Technologies were disseminated on biopesticide (Trichogramma), oyster mushroom production and vermicomposting to recycle the spent waste of oyster mushroom. Extension material brought out in the local language was distributed to the villagers.

## Aquaculture

A project was undertaken on enhancing livelihood of SC/ST people in Orissa and Tamil Nadu by MS Swaminathan Research Foundation, Chennai for introducing various aquaculture enterprises to 250 participants in backyard ornamental fish culture and carp seed production. Fingerlings produced were distributed to the entrepreneurs for stocking. Various aspects on ornamental fish breeding, carp nursery and gold fish culture were disseminated to the self help groups, which helped them in earning a revenue of ₹ 200/month/household and an income of about ₹ 3200/from carp nursery and ₹ 8500/from vegetables and flowers as bund crops. Community ponds were effectively utilized by stocking with rohu, catla, mirgali, Chinese & Indian carps and marketable size fish was harvested. Technologies transferred to the entrepreneurs on community pond aquaculture, carp nursery and ornamental fish culture and extension material in local language were distributed to the beneficiaries.

A project on utilization of integrated marine living resources by SC youth from Tuticorin region of Tamil Nadu was undertaken by College of Fisheries, Tuticorin. Lobster fattening and crab culture techniques were standardized and demonstrated to the beneficiaries, who have grown lobsters in cages for about 4 months and produced 250-350gms size for marketing. The beneficiaries were helped in availing credit facilities and bank loan. In the project around 120 people of target population have been trained. A manual on crab and lobster fattening technology was brought out for the distribution to the beneficiaries.

A project on freshwater prawn culture was implemented in village Naubasta Kalan in Allahabad District by University of Allahabad. Demonstrations were conducted on prawn hatchery technology and standardized methods for biocontrol of prawn disease for integrated giant freshwater prawn (Macrobrachium rosenbergii) culture. Trainings were conducted on scampi culture, breeding and seed production and microbial inputs were utilized for feed silage preparation using indigenous technology. Apiculture activity was introduced and disseminated to the target population and extension material was distributed to the beneficiaries.

## Genetic Counseling

Tribal population of northeast region was screened for anaemia and nutritional deficiencies through a programme implemented at Vivekananda Institute of Medical Sciences, Kolkata. The tribes of Arunachal Pradesh, Assam and Tripura displayed high incidence of anaemia. Awareness programme was organized for school children on inherited anaemia in pre-marital and pre-natal population. More than 1500 families from 80 villages were educated on anaemia, thalassaemia and hemoglobinopathies and counseling was provided to them. Leaflets were distributed for awareness of anaemia with special reference to Thalassaemia in local languages.

A project on molecular epideomiology of alcoholism was implemented jointly by Centre for Cellular and Molecular Biology, Hyderabad and Tribal Research Centre, Udagamandalam, Nilgiri District to study Kota tribe population in Nilgiri hills. General health awareness camps were arranged in seven villages for periodical medical check up. Blood samples were collected from 620 individuals and clinical data on health profile revealed that the disease morbidity is more for hypertension, heart related problems and diabetes.

Genealogies indicated that the alcohol drinking habit is running in the families for the past two to three generations. Study undertaken on the health awareness camps and medical camps were found beneficial in counselling the tribal population.

## Floriculture

Over 180 women from Shillong, Itanagar and Chandigarh were trained in cultivation of commercially important and hybrid species of orchids. A total of 118 women of Shillong have raised 6 species *viz.* Dendrobium fimbriatum var. oculatum, D. longicornu, D.lituiflorum, C. pendulum, C. aloifolium and C. giganteum in their backyard. These plants have started blooming. The beneficiaries have started selling their harvest in local market. Through S&T Council, Bhopal 200 women were trained in cultivation of gladiolus, marigold, mogra, rose and tuberose.

## Food Processing

At College of Horticulture, Vellanikara, Thrissur, 88 women were trained in various aspects of entrepreneurship in preparation of cocoa products including molding, packaging and marketing. Twenty-two women have already set up their production unit and are selling their products in local market. In another project at Fisheries College, Tuticorin, 500 women were trained in cook-chill technology for seafood and various parameters for quality assurance *viz.* estimation of moisture content, bacteria load, presence of bacterial species, the nutritive value, shelf-life etc. The trained women were grouped in 48 SHGs for establishing market for seafood products.

At the Central Food Technological Research Institute, Mysore data on biochemical and microbial analyses for nutritionally important factors like protein, amino acids, isoflavones, genistein, diadzein and methods for improvement of shelf life were collected on for 'Tofu' preparation. Also a byproduct of Tofu preparation, Okara, was found to be rich in protein and amino acids has been used as a food supplement. In another project at CFTRI, 17 women were trained in preparation of various food products of high nutritive value, methods of removal of anti-nutritional factors like phytic acid and trypsin inhibitor found in soya bean, better texture and improved shelf life. The food items included fermented products of soya bean and fruit pulp. Also 34 women were trained in spirulina production and its application as health food. Through CFTRI, technologies for rural bakery are being disseminated to rural women in three villages.

## Poultry and livestock farming

The 'Action Research on Health and Socio Economic Development', Bolangir, Orissa, identified 100 women and trained them in poultry rearing. The beneficiaries were provided with parent birds of Carigold (dual purpose) and Vanaraja varieties. The beneficiaries were able to earn ₹ 1000/per month through the sale of eggs and birds. The beneficiaries were also trained in feed formulation using local ingredients, management practices and entrepreneurship skill. The trained women formed SHGs and are approaching a bank for setting up their poultry farms. The market is established through state owned apex cooperative known as OPOLFED.

Through three projects supported at Tamil Nadu Veterinary & Agricultural Sciences University and its Extension Centres at Namakkal, Chennai and Tiruchirapally, over 1200 women were trained in poultry farming including feed formulation, farm management / health care and preparation of various food items such as quail egg pickle and ready-to-eat food products etc.

*Aquaculture*: Under a project at the North Eastern Hill University, Shillong, techniques were standardized for induced breeding for common carp and 4,00,000 fish seeds were reared up-to fries for distribution to the beneficiaries. Selected 92 women were trained in fish farming techniques including water quality testing, pond maintenance, induced breeding, harvesting and feed formulation using slaughter house waste. They were provided with fish fries for rearing in their own pond.

Over 470 women from Bhopal, Shimla, Solan and Bhubaneshwar were trained in polyculture fish farming. Interested women were also provided advance training in composite carp culture, integrated fish farming with duck/poultry, pond management and value added products including fish processing. These women are selling their products in local market and are earning ₹ 3000/per harvest/family.

Considering the opportunities in ornamental fish, a project was supported at the University of North Bengal. The identified women, 57 from Darjeeling and 54 from Jalpaigudi district were trained in rearing and breeding of guppy (Poecilia reticulate) in cemented cisterns and angel (Pterophyllum scalaer) in ponds. The women are being trained in management skill and marketing.

## Organic farming / vermicompost

At the Sri Sathya Sai Institute of Higher Learning, Anantapur Campus, Andhra Pradesh, neem varieties with higher level of azadirachtin were identified in 5 villages.

Thirty trees were selected and 40 women from two villages were trained in fruit collection, drying and storage. Specialised training on maintenance of neem processing gadgets, processing of neem fruits for production of cake, oil or powder, maintenance of aza-rich plantlets, hardening, harvesting and application of neem based biopesticide was provided to five resource persons At the University of Agricultural Sciences, GKVK Campus, Bangalore, five villages of Bangalore and Kolar were surveyed for the availability of organic waste.

The waste materials were mostly sugarcane trash, coconut coir pith and vegetable wastes. Demonstration programmes for 20 to 25 farmwomen from each identified village were organized. Six interested women were given onfarm training.

The technique of Trichoderma enriched vermicomposting was perfected and demonstrated at University campus.

At MSSRF, Chennai, phosphate-solubilising local strains were collected and screened. Simultaneously several awareness programmes on application of biofertilizer were organized for the villagers of Rediarchatram block. About 300 interested farmers were trained in large-scale production of biofertilizer. Under the project support at Dangoria Charitable Trust Hyderabad, apart from training women in processing various items of food, 100 women from 13 villages were trained in production of vermicompost, neem-based biopesticide and their application in organic farming, About 100 vermicompost units were set up using the ring method. The beneficiaries produced over 130 quintals of compost. The beneficiaries were encouraged to raise nurseries of drumstick, papaya and creeper spinach at their backyard using vermicompost. The beneficiaries were able to sell excess vermicompost and planting material.

## Horticulture

Through the Tamil Nadu Agricultural University, Coimbatore, 54 progressive farm women were identified for training on cultivation of hybrid rice. The women were provided with quality seed their cultivation over an area of approximately four acre. The yield for the first harvest (December to May) was recorded at 2000 kg/ha. The beneficiaries were able to earn an additional income of ₹.20,000/per acre/year through three crops in an year.

At Supi farms, Uttaranchal non-traditional vegetable crops were grown on farmers' land in a small way. An herbal garden was established. Incidence of pests to various crops was evaluated and farmers were trained on production of vermicompost and application of neem based biopesticide in horticultural crops.

## Mushroom cultivation

In an attempt to empower rural women through transfer of mushroom cultivation technology 93, women were trained at Himachal Pradesh Krishi Pradesh Vishwavidyalaya, Palampur and 750 at Tropical Botanic Garden & Research Institute, Kerala in mushroom cultivation. In addition, 15 women were trained in spawn production at TBGRI. The beneficiaries were given 15 days training in cultivation of Oyster and Agaricus species using wheat straw.

The first harvest of about 1000 kg was sold @ ₹ 16/kg. The beneficiaries were also trained in various recipies of mushroom. Market linkage has been established for their products. Under the project support at Dangoria Charitable Trust Hyderabad, apart from training women in processing various items of food, 100 women from 13 villages were trained in production of vermicompost, neem-based biopesticide and their application in organic farming, About 100 vermicompost units were set up using the ring method. The beneficiaries produced over 130 quintals of compost.

## Sericulture

A total of 550 women were trained for various activities including the organic cultivation of mulberry & rearing of cocoon and value added products from waste at the Central Sericultural Research & Training Institute, Mysore and at Vidyasagar University, Midnapore, West Bengal. The beneficiaries are getting an additional income through the sale of value added products and surplus vermicompost.

## Healthcare

Kidwai Memorial Institute of Oncology, Bangalore, Karnataka organized over 50 awareness camps on Carcinoma cervix to educate 2071 women for its prevention. Blood samples collected from 1830 women based on their medical history, 1812 were found negative and 18 positive cases were referred for treatment. A project on mental retardation including awareness generation among rual women, screening for early detection and genetic counseling was supported to Manovikas Bio-Medical Research. Over 7,000 families from three Gram Panchayats were surveyed. Ninety-one of these showing signs of mental retardation were shortlisted. The families were informed about the possible risk factors in future. Analysis of blood samples from these for genetic carrier showed only 4 cases of low level of fragile X chromosomes.

A project is supported at Saraswati Dental College, Lucknow for dental fluorosis. Water samples from 12 villages around Lucknow and Kanpur were analysed for fluoride content. Health camps were organized in 4 villages with a team of dental surgeon. A total of 2860 people were examined for dental fluorosis.

The severely affected persons were advised for regular dental examination at SGPGI, Lucknow. They were also provided with calcium and cholicalciferol supplement.

The blood samples from the severely affected boys and girls within the age group of 15 years and pregnant women were examined for their blood calcium level. Under another project supported at Mahavir Hospital And Research Centre, Hyderabad, Andhra Pradesh, blood lead levels in children and adults including men and women were measured to study environmental genotoxicity in women and children occupationally exposed to lead, a heavy metal.

# 8

# Agricultural Biotechnology and Forestry

Unlike crops, only a small percentage of the current world forest plantings comprise material resulting from genetic improvement of any type. Genetic improvement of forest species has essentially been limited to selection and multiplication of well-adapted genotypes, although for industrial forestry species, especially Eucalyptus and Pinus, recurrent selection for population improvement has been widely adopted.

Reproductive biology and patterns of variation of these species is reasonably well documented. Vigour, form and wood quality are the priority selection criteria. As regards non-industrial tree species, the knowledge of biology and patterns of variations are poorly understood and selection criteria are also diffused. The breeding work is essentially confined to species testing, the assessment of the biological features and genetic conservation activities. The use of biotechnologies to the various tree species would thus vary according to the current and potential use of and level of previous genetic improvement work done in the species.

Gene structure and function. Trees must often cope with environmental stress to survive. Anaerobic stress, caused by a lack of oxygen, occurs whenever demand for oxygen exceeds supply. Although a plant may experience anaerobic stress after a flood, it may also occur within metabolically active tissues of a nonflooded plant. Anaerobic stress activates a small set of genes and alters biochemical processes that generate cellular energy. Among the genes activated by anaerobic stress, those encoding alcohol dehydrogenase (ADH) are understood the best. ADH is the enzyme that catalyzes the interconversion of acetaldehyde and ethanol. This enzyme is essential for plants to survive short-term exposures to flooding, but its physiological function is not clearly understood.

Patterns of ADH expression in woody plants are unusual when compared to nonwoody plants, probably because anaerobic stress commonly occurs within metabolically active tissues of tree stems, such as the cambium. Using classical genetic tools and recombinant DNA methods, we are studying the gene (or genes) responsible for ADH expression in different tissues of pines and cottonwoods to understand the molecular mechanisms responsible for switching these genes on and off. These studies will enable us to better understand how anaerobic stress affects tree growth and wood production.

Tissue culture and plant regeneration. Moving genes from test tubes into actual trees requires three methods: one method to shuttle genes into plant cells, one to aseptically grow plant tissues, and one to regenerate trees from isolated tissues. Our research addresses several of these areas.

One project regenerates plants from selected clones of the eastern cottonwood (Populus deltoides) and from immature Populus embryos. Because they grow so rapidly, Populus species are prime candidates for windbreaks and for producing wood or fiber. Four-to ten-year-old Populus trees can be harvested for woody biomass (to be burned directly or converted to chemical fuels) or for pulpwood. We can dramatically alter plant development in culture by changing the type of sugar or hormones contained in growth media. We can now regenerate plants from isolated tissues of mature individuals as well as from immature embryos. Because hybrid embryos can be rescued before they abort, the latter method allows sexually incompatible Populus species to be crossed.

In collaboration with researchers from the Department of Plant Pathology, another project transfers genes into trees and shrubs that form root nodules containing Frankia, a nitrogen-fixing bacterium with a diverse range of plant hosts. Initial experiments have infected plant tissues from red alder (Alnus rubra) and Russian olive (Elaeagnus angustifolia) with the bacterium Agrobacterium rhizogenes. By infecting these and other plant species, Agrobacterium transfers a few bacterial genes into the cells of the host plants. This natural process provides a mechanism to shuttle selected genes into plants. From infected plants, we have observed morphologically and biochemically distinct shoots and roots we believe to contain transferred genes. These methods will enable genetic manipulation of nitrogen fixation, which may reduce the need for nitrogen fertilizers.

## In-vitro Culture Techniques

Successful "slow growth" storage of cultured tissues and organs has been reported for several forest species. The major bottleneck in several cases, however, is the regeneration of plants from the cryopreserved materials. In vitro conservation can be effective for threatened gene pools. The main advantage of cryopreservation would be the maintenance of the juvenile state to facilitate capture of genetic gains through clonal forestry with industrial species and a few other species for which good breeding programmes are already in progress. In vitro cultured materials would also find favour for germplasm transfer of certain species.

In vitro rejuvenation or accelerated maturation has been obtained in a few forest species and could be exploited for clonal forestry and accelerated breeding cycles, respectively. Micropropagation protocols are known for more than 100 forest species, but commercial exploitation of the technique lags behind because of paucity of data economics of the vitro plantations. Nonetheless, the technique is being used for micropropagation of outstanding clones in breeding programmes, i.e. of tropical eucalypts and for early multiplication of selected varieties prior to release. The efficacy and efficiency of somatic embryogenesis and "artificial seed" is being analysed and improved both in selected public and private institutions and the approach holds promise under certain circumstances. Cryopreservation of somatic embryos with field evaluations of emblings and scaled-up bioreactor-based multiplication of selected cryopreserved embryos is expected to be commercialized for several tree species within the next five to ten years (Gupta et al., 1993).

In agroforestry or community forestry programmes where intensive cropping of forest species is undertaken, micropropagation of desired clones assumes still greater economic importance. Eucalyptus plantations are one of the leading forestry systems. About ten years ago, Gupta and colleagues at the National Chemical Laboratory (NCL), Pune, India, spotted extremely promising trees of Eucalyptus camaldulensis and E. torelliana in the forests of South India and developed micropropagation techniques for their multiplication. Some 50 000 plantlets of E. torelliana and 20 000 plantlets of E. camaldulensis could be produced within a year from just one nodal segment taken from an adult tree of each species. More work is needed on the standardization of micropropagation techniques to render them more cost effective.

Some of the developing Asian countries have made considerable progress in exploiting biotechnology for forestry. For instance, Indian scientists at the NCL were the first to standardize in vitro propagation of 100-year old teak (Tectona grandis), and 30-year old Eucalyptus citriodora and other Eucalyptus species. Pilot projects on teak, eucalypts, bamboo, sandal and salvadora plantations based on in vitro cultured materials are in progress and it is hoped that micropropagation of other useful species such as Tamarindus indica, Azadirachta indica, Santalum album, and Morus alba will also be standardized soon.

An interesting success of the Indian scientists was in vitro flowering of bamboo-a species that flowers once in a decade or more, which is a significant development for bamboo breeding. Indian scientists have succeeded also in standardizing the technique for several nitrogen-fixing tree species such as Acacia nilotica, Albizia lebbek, Dalbergia latifolia, D. sissoo, Leucaena leucocephala, Prosopis cineraria, Sesbania grandiflora, and S. sesban. The work on nitrogen-fixing species is particularly significant from the point of view of nitrogen economy of the soil. Selection of efficient strains of Rhizobium, Azotobacter and Mycorrhizae should also be emphasized concomitantly.

In China, during the past 15 years or so, useful work on the culture of anther, shoot, embryo, endosperm, ovary, suspension cell and protoplast of woody species has been done. In vitro culture plantlets were reported in about 100 woody species, including poplar, eucalypts, paulownia, and Chinese fir. In vitro culture techniques are being used for developing fast-growing, disease-resistant, and proficient nitrogen-fixing Casuarinas for establishing new shelterbelts and rehabilitating old ones.

Somatic embryogenesis, which allows automation of micropropagation and renders it highly cost effective, can play an important role in reforestation. So far, somatic embryogenesis has been reported in at least 60 species of woody trees belonging to 44 genera and 25 families. In some cases the process has been patented by private companies for commercial exploitation. Greater research and development efforts are needed to standardize the technique in developing countries for large-scale reforestation. Franclet (1989) observed rather striking homogenety in progenies of certain distant hybrids of Eucalyptus. The uniformity was postulated to be due to apomictic seed setting in the F, plants. This needs to be confirmed and if found true, the finding will revolutionize clonal propagation of hybrid and other populations of Eucalyptus.

Pollen plantlets from about 30 woody species have been obtained in China. Indian scientists also produced haploids of a few tree species, including Albizia lebbeck. The haploids were used by Chinese scientists to produce homozygotes and further breeding work. Given the long life cycle of trees, haploid technology is particularly suitable for breeding improved varieties of forest species. Somaclonal selection has recently been used to produce and select clones of hybrid poplar resistant to Septoria musiva (Ostry and Skilling, 1988). Exposure of cell cultures to disease toxins, saline/alkaline conditions, unusual temperature regimes, etc. facilitate isolation of resistant types. Using this technique, Eucalyptus camaldulensis clones resistant to saline soils were developed. However, no somaclonal tree has yet been put to commercial use. The value of hybrid populations in terms of growth, vigour and productivity of tree species, including the Eucalyptus hybrid populations described earlier, is well-known. Controlled hybridization of selected parents holds great promise for producing novel and productive populations.

For those species that cannot be hybridized sexually, somatic hybridization-protoplast fusion and regeneration of plantlets from "hybrid" protoplasm is a possible alternative. As regards forest and other woody species, protoplast fusion and regeneration have been achieved only in a few species, such as elms, poplar and rubber. Concerted effort is needed to standardize the protoplast culture, fusion and regeneration technology for important woody species. RFLPs and RAPDs are being used extensively for quantification of genetic variation, markerassisted selection and genotype verification in several forest species. Transgenic eucalypts are being produced for incorporating cold tolerance to enable expansion of this species into cold-prone zones to which this species has little tolerance.

Another objective is to transfer pest resistance in poplars and some tropical hardwoods, while a further attractive proposition is the reduction of lignin biosynthesis in the pulp species. A recent FAO study (1993) carried out by Haines under the André Mayer Research Fellowship underlines that the use of biotechnologies for the improvement of tree species necessitates intimate knowledge of biology, breeding systems and variations in the pattern of the target species, which is generally poor. While new technologies could be effective and expedient in specific cases, major emphasis on the improvement of most tree species will in the near future remain on taxonomic studies of variation, species and provenance testing, field assessment and genetic conservation. A major constraint in the use of biotechnology in tree improvement programmes is the lack of skilled tree breeders.

The study concludes: "In the short term, for industrial species, with known biology and variation patterns, opportunities exist for technology replacement in the incorporation of micropropagation into integrated multiplication systems for intensively bred species, and in the use of molecular markers in genotype verification. For non-industrial species and for presently less well-known hardwoods with potential for industrial use, opportunities exist for the use of markers in studies of genetic variation and breeding systems. In the longer term, significant applications of new technologies are likely. Major research priorities include the acquisition of better understanding of molecular genetic processes in forest trees, particularly those underlying

having close links or interaction with public research institutes as well as universities and thus faces a broad set of different policy interventions by the government underlying the National Innovation System (NIS).

The National System of Innovation (NSI) perspective is a convenient framework to understand the process of innovation occurring in an economy and especially within the manufacturing sector. It focuses on the country-specific factors influencing the process of innovation and economic growth. The NSI approach has evolved over the years both analytically and empirically with the contributions from Freeman, Lundvall and Nelson, Edquist (1997, 2002). NSI has became popular as an analytical framework to study innovation at the national, regional and even Sectoral level.

Though there are several definitions of NSI but that best fit in to the present context is that of Metcalfe (1995). NSI is defined by Metcalfe (1995) as a "set of distinct institutions which jointly and individually contribute to the development and diffusion of new technologies and which provides the framework within which governments form and implement policies to influence the innovation process. As such it is a system of interconnected institutions to create, store and transfer the knowledge, skills and artefacts which define new technologies".

NSI framework is comprised of various components such as the government, independent research institutes, firms and higher education system etc and their interrelationships. Innovation or government policy instruments and institutions play a very important role in cementing the relationship among the various components for the success of NSI. An application of this framework shall help in analysing the systemic failure that hamper the generation of innovations. Public policy can be applied to correct such failure.

There are several NSI studies on Biotechnology in the context of the developed world but the studies in the developing countries are rather limited. Biotechnology NSI in India has evolved over time and has been instrumental in creating infrastructure and capabilities in promoting biotechnology innovation in the country. There are 241 institutions and 321 firms engaged in various dimension of biotechnology in the country. Indian Biotechnology industry is now among the top 12 countries in the world and like its global counterpart, is dominated by the healthcare sector. The recently released "Draft National Biotechnology Strategy 2005" by the Indian Government is a step forward in the direction of promotion and development of biotechnology to withstand global challenges and yet, solve the local healthcare needs of the country. Studies on biotechnology in India by and a recent empirical analysis of the health innovation system of seven developing countries further supports the role of the Governments in providing the sustainable environment for the successful development of health biotechnologies.

According to Acharya *et. al.* government in developing countries is one of the key actors in prioritizing, implementing new technologies and promoting innovation to address their specific health and development needs.

The governments of the countries where biotechnology innovation system has been successful in perform or support the following functions/activities such as knowledge support, commercialization of research, collaborative research, financial support, regulation, IPR and resolving ethical issues etc.

Thus it would be interesting to analyse the role of the government in the promotion of the biotechnology in the country with special focus on the healthcare sector.

Hence, the present study traces the role of government or national approaches in terms of policies, institutions and instruments in the promotion of biotechnology innovation system in India with a special focus on the healthcare biotechnology sector. A systemic framework similar to the EPHOITE (2003) project is being adapted to examine the direction of public policies, institutions and instruments in addressing the issues of knowledge support, commercialization of research, collaborative research, financial support, regulation, IPR and ethical issues in the area of healthcare biotechnology in Indian context.

The significant features of national biotechnology innovation system with focus on healthcare biotechnology with the help of various S&T indicators. The various government initiatives in terms of various policies, institutions and instruments in the framework of NSI while the lessons learnt along with the policy implications and challenges for the growth of healthcare biotechnology innovation.

## BIOTECHNOLOGY AND INDIA AGRICULTURE

We need to recongnise that, just as the food reequirements of today's population of nearly 6 billion people could not have been met by the technologies of the 1940's, we cannot assume that current practices will feed the population of 8 billion expected by 2020. New approaches are needed in addition to the continued improvement of existing methods of crop and animal husbandry and food processing.

The strategic integration of biotechnology tools into India agricultural systems can revolutionise Indian farming and usher in a new era. Genetic engineering is clearly the most revolutionary tool to impact agricultural research since the discovery of genetics by Mendel. Prior to genetic engineering, the exchange of DNA material was possible only between individual organisms of the same species. With the advent of genetic engineering in 1972, scientists have been able to identify specific genes associated with desirable traits in one organism. For example, a gene from bacteria, virus or animals may be transferred into plants to produce genetically modified plants having changed characteristics. This method therefore allows mixing of the genetic material from species that cannot otherwise breed naturally.

Genetic engineering can be vital for an agrarian country like India. It can help in minimising the crop damage through disease and pest resistant varieties, reducing the use of chemicals, enhancing stress tolerance in crop plants, thus permitting productive farming on unproductive lands, etc. One could even extend the growing season of crops and minimise losses due to environmental factors on the one hand and increase the shelf life of fruits and vegetables, on the other, thus minimising losses due to food spoilage.

We can expand the market vista and improve food quality. Biotechnology can also produce plants that possess healthy fats and oils, possess increased nutritive value, and create a whole range of higher value feeds. Biotechnology has even the power of producing biodegradable plastics, edible vaccines, etc.

It is only through the blending of the 'gene revolution' with our experience in the 'geen revolution' that we can reach our goal of 'evergreen revolution' and also 'nutritional revolution'.

The advantage of the gene revolution is that it is relatively scale neutral, benefiting big and small farmers alike, It is also environment friendly.

Thus, it can be of great help to the smallest farmer with limited resources in increasing farm productivity through the availability of improved but powerful seed. It can also reduce a farmer's dependency on chemical inputs such as pesticides and fertilizers.

Modern biotechnology offers unlimited opportunity for enhancing genetic potential of crops and other commodities, management of biotic and abiotic stresses, bio-remediation and organic recycling. India is one of the main centres of agricultural biodiversity and its gene richness can greatly complement the developments in modern biotechnology. Likewise, new developments in GIS, remote sensing, and crop modelling, provide new opportunities for integrated management of natural resources. The revolution in informatics provides opportunities for sharing latest information for research and planning in a highly organised and efficient manner. With the globalisation of economy, the opportunities for value addition and post harvest management are also immense.

With clever blending of technologies, the India farmer can usher in new era of confidence and performance. Sir Francis Bacon had once said: *'It would be an unsound fancy to expect that things which have never yet been done, can be done except by methods which have never been tried'* In the new context, we have to review the new role of biotechnology. Also, it is in this context that we have to view the rapid advances that modern biotechnology is making, including the area of transgenics. Let us focus on some of the developments in this area, since considerable interest and controversy has been created around the world today in this exciting area.

## Movement in Transgenics

The advent of genetic modification over the last two decades has enabled plant breeders to develop new varieties of crops at a faster rate than was possible using traditional methods, with huge potential for further beneficial development.

Tobacco was the first plant to be genetically transformed\, in 1983, with cereals beginning in 1990, but it is only recently that products such as GM soya have reached the market place in Europe, and, for a variety of reasons, given rise to concern & controversy.

The concern must be addressed seriously if society is to exploit the new technologies appropriately. We will address this issue rather extensively a little later. Whereas in India, we have been discussing and debating the issues, the rest of the world is galloping ahead.

*Table. Ag Biotech Product Sales*

| $ Millions | | Sales | | Annual Growth Rate | |
|---|---|---|---|---|---|
| | 1994 | 1997 | 2002 | 1994-97 | 1997-02 |
| Transgenic seeds and plants | 20 | 405 | 2,030 | 173% | 38% |
| Animal growth horomes | 80 | 225 | 405 | 31 | 12 |
| Biopesticides | 35 | 65 | 110 | 23 | 11 |
| Other | 107 | 180 | 340 | 19 | 14 |
| Total U.S. Sales | $242 | $875 | $2,885 | 53% | 27% |

Development, field testing, and commercialisation of transgenic crops, is now recongnized by the international scientific community to be an essential strategy for food security. Fortyfive countries, including India, have conducted transgenic crop field trials up to 1997. The activity in the developing countries has been shown in Table below.

*Table. Releases of Transgenic Plants in Developing Countries (by Species and Introduced Trait)*

| Species | Introduced Trait | | | | | |
|---|---|---|---|---|---|---|
| | Total Fied releases | Herbicide Resistance | Insect Resistance | Virus Resistance | Product quality | Others |
| Maize | 46 | 29 | 16 | 1 | 3 | 5 |
| Soya bean | 27 | 25 | — | 1 | 1 | — |
| Cotton | 24 | 16 | 15 | — | — | — |
| Tomato | 19 | — | 2 | 1 | 16 | — |
| Potato | 13 | — | 1 | 6 | — | 6 |
| Subtotal | 129 | 70 | 34 | 9 | 20 | 11 |
| Other species | 30 | | | | | |
| Total | 159 | | | | | |

The emphasis has been mainly on traits for herbicide, insect, virus & fungal resistance, and on product quality with a limited number of genes. Out of a total 159 releases. Argentina leads the way with 43, followed by chile, Mexico, Puerto Rico and the Republic of South Africa, with 17-20 each. India is far behind with just a few releases. There are fears relating to transgenic plants in society. These need to be addressed, not by brushing them aside, but by careful scientific analysis and educational programmes. All biosafety protocols must be followed rigidly. Let us deal with this issue more exhaustively now.

## Public Concerns

The potential of biotechnology as a method to enhance agriculture productivity in the future has been accepted globally. However, because of its revolutionary nature, risk and uncertainty may be created by the process of genetic engineering and by the resulting genetically modified products. This may result from the introduction of a new unrelated DNA sequence into a recipient organism.

The introduced DNA may have some unexpected effects on the cellular processes of the recipient organism. In addition to this, introduction of antibiotic resistance genes, as selection markers, pose serious implications in public health especially in genetically modified plants that are directly used for food purposes.

Risks are also associated with the genetically modified plants that are released into the environment. The nature of interactions with other organisms of the natural ecosystems cannot be anticipated without proper scientific testing. For example, modified plants with enhanced resistance to pests or disease threaten to transfer resistance to the wild relatives, which will have implications for biodiversity and ecosystem integrity. These, and many more, doubts plague the minds of common people. Therefore, it is very clear that if the biotechnology potential has to be completely realized, it has to be done in an extremely responsible manner. At each step, proper testing and scientific data have to provided. Not only this, proper regulatory and policy mechanisms have to be put into place.

It is therefore, absolutely essential, that conscious efforts be made especially by the developing countries to initiate well-defined programmes for the development and regulations of genetically modified plants. To make this possible and to benefically, ethically and sustainably reap the benefits of biotechnology, there is a pressing need for scientists, researchers, policy makers, NGOS, progressive farmers, industrialists, and representatives of the government, to come together on a common forum in order to discuss the common concerns and find solutions to them.

Will modified plants transfer their introduced genes into wild relatives growing nearby? Will modified plants that produce new compounds disrupt the 'balance of nature' in some way? Could the planting of a restricted number of cultivators leave crop plants more susceptible to disease? Could the planting of a restricted number of cultivators lead to a reduction in biodiversity (of crop plants, weed species, insects and microflora in the fields in question)?

Will genetically modified plants be able to avoid the factors that regulate natural populations and thereby change the usual ' balance' between populations? Similarly, there are concerns about the social and economic effects. For example, some of these concerns are:

- How will the structure of farming (particularly in developing countries) be affected by biotechnology?
- How will patent laws affect traditional breeders 'rights (*e.g.* the right to save seed from one year to the next)?
- Will plant breeding be left increasingly in hands of a few companies, and if so, what effects might this have?
- Will some countries be 'plundered' for their genetic resources?
- We also have a set of ethical and moral issues that need addressing. In particular,
- The consumer has the right. What will consumers be told about the new food products?
- Is it acceptable to 'interfere' with nature through genetic engineering?

- Do we have the right not to use all means available to improve crop plants, especially when so many people are under or malnourished?

Then there are several regulatory issues. Among these, the three key issues are:

- Do current regulations give sufficient protection to farmers, consumers, those who have invested in, and those engaged in research?
- Is there sufficient international legislation to ensure environmental protection?
- Do current regulations compromise the competitiveness of biotechnology companies by being excessively restrictive?

The Royal society of London had appointed a group of experts to examine various aspects including the scientific evidence concerning the risk of transfer of genes from Genetically Modified (GM) crop plants to wild species and non-GM crops, the uptake of genes from GM food by the digestive system, and the current state of the regulatory system.

The experts concluded that the chances of gene transfer happening are slight, provided the regulatory processes are followed, but that this must be kept under consideration.

The experts also concluded that the uptake of genes via the food chain is not a new issue because genes (*i.e.* DNA) are normal constituents of the human diet. Many products from GM Plants, such as sugar prepared from GM tomato paste, are so similar that they are regarded as 'substantially equivalent' others, for example flour from GM soya, may contain a new gene or its product, although many of the purification processes involved in food production will destroy and DNA present in the raw material.

Some GM foods have been produced using an antibiotic resistance 'market' gene, which is a laboratory device designed to identify genetically transformed plants.

The Royal Society has shown concern about the use of such genes in food products and it suggested that any further increase in the use of such markers in the human or animal food chain would be undersirable. Gm crop plants have been produced to improve insect tolerance and virus resistance, and to include herbicide tolerance, so that transferred to non-target species of plants, and that the development of resistance by target pests is minimised, the regulatory authorities must be assured that any negative effects would be no greater than those resulting from conventional procedures; and that any long term effects on the environment and ecology would be closely monitored, with statutory restrictions in place to control marketing; and also that best practice advice was adopted by growers. The reliance on a case by case approach to the legislation, may result in lack of analysis of the overall impact of the technology on agriculture and the environment, and of the long term effects of gm. Although mechanisms are already in place to regulate many individual aspects of GM technology, there is no means for looking at GM technology as a whole. As independent, overarching, regulatory body is needed to span departmental responsibilities, monitor the enforcement of existing or future regulations, and strengthen the guidelines to growers of GM crops, such as those specifying the isolation distances between Gm and non-GM crops. The body should also review and monitor the membership of advisory committees and regulatory bodies.

## INSTITUTIONAL SETUP ANDBUDGETARY ALLOCATIONS IN INDIA

India is one of the first few countries, among the developing countries, to have recognized the importance of biotechnology as a tool to advance growth of agricultural and health sectors as early as in 1980s. India's Sixth Five Year Plan (1980-85) was the first policy document to cover biotechnology development in the country.

The plan document proposed to strengthen and develop capabilities in areas such as immunology, genetics, communicable diseases, etc. In this context, referring to the Council of Scientific and Industrial Research (CSIR), the document suggested to ensure coordination on inter-institutional, inter-agency and on multi-disciplinary basis, full utilization of existing facilities and infrastructures in major areas including biotechnology.

Programmes in the area of biotechnology included, tissue culture application for medicinal and economic plants, fermentation technology and enzyme engineering for chemicals, antibiotics and other medical product development; agricultural and forest residues and slaughterhouse wastes utilization and emerging areas like genetic engineering and molecular biology.

The existing national laboratories under the S& T agencies, such as Indian Council of Medical esearch (ICMR) and Council for Scientific and Industrial Research (CSIR) had initiated several esearch programmes to fulfill the above plan objectives. At the top, an apex official agency *viz.* ational Biotechnology Board (NBTB) was set up in 1982, to spearhead development of biotechnology. The NBTB was chaired by Member (Science) of the Indian Planning ommission and had representation of almost all the S&T agencies in the country *viz.* epartment of Science and Technology (DST), Council for Scientific and Industrial

Research CSIR), Indian Council of Agricultural Research (ICAR), Indian Council for Medical Research ICMR), Department of Atomic Energy (DAE) and the University Grants Commission (UGC).

NBTB was formed with the specific purpose of the identification of priority areas and for volving a long-term plan for the country in biotechnology as well as to initiate and promote uch activities as conducive for further development of various areas in biotechnology. The BTB issued the " Long Term Plan in Biotechnology for India" in April 1983. This document pelt out priorities for biotechnology in India in view of the national objectives such as self ufficiency in food, clothing and housing, adequate health and hygiene, provision of adequate nergy and transportation, protection of environment, gainful employment, industrial growth and alance in international trade. Later in 1986, NBTB graduated to a full-fledged government epartment called Department of Biotechnology.

At present in India, there are six major agencies responsible for financing and supporting esearch in the realm of biotechnology apart from other sciences. They are Department of cience and Technology (DST), Department of Biotechnology (DBT), Council of Scientific and ndustrial Research (CSIR), Indian Council of Medical Research (ICMR), Indian Council of griculture Research (ICAR) and University Grants Commission (UGC), Department of cientific and Industrial Research (DSIR). DST, DBT and DSIR are part of Ministry of Science nd Technology while ICMR is with Ministry of Health, ICAR with Ministry of Agriculture and UGC with Ministry of Human Resource and Development. DSIR is the funding agency for CSIR and both of them independently fund biotechnology related research programmes.

Allocations for all of these agencies have gone up in the last decade. Out of this, DBT is the only agency completely devoted to R&D in biotechnology. It is very difficult to estimate the total allocations for this sector per se from other aforementioned agencies as in some cases the allocations are not separately marked as allocations for biotechnology. One faces this kind of constraint especially with those organisations, which are focusing on technological solutions and are not committed for X or Y nature of technology.

Thus separately accounting for biotechnology becomes difficult. It would probably become possible only if a detailed survey at the institutional level is undertaken. A broad idea about total allocations by major ndian funding agencies to Science and technology related projects and not necessarily to biotechnology alone. In case of UGC it gives a broad idea not only about S&T related projects but also about other research streams.

## Department of Biotechnology (DBT)

There has been a significant increase in Government of India's outlays for biotechnology over the past decade. Since the time of establishment, in 1986, the allocation for the Department has increased manifold. The budgetary allocations have gone up from ₹ 404 million in 1987-88 to ₹ 1,138 million in 1997-98 and by 2001-2002 it became ₹ 1,863 million. Though the current price allocation figures may not give a complete picture but the budgeted figure for 2002-03 shows a doubling of allocation to ₹ 2,356 million. In India the developmental allocations are generally made for five years under the National Five Year Plans.

The Government has recently finalized the Tenth Five Year Plan (2002-2007). The Working Group on Science and technology for the Tenth Five Year Plan, constituted by the Planning Commission, has proposed an outlay of ₹ 20,750 million for the period 2002 to 2007. This marks a sharp increase of 234 per cent from the budgetary provisions made during the Ninth Plan period (1997-2002) which totaled at ₹ 6,215 million only. The Vision Statement for the Tenth Five Year Plan enumerates the proposal for human genome sequences, proteomics, structural biology and bioinformatics.

The DBT has taken special precaution to find crucial balance between different sections of society as far as technology absorption is concerned apart from promoting industrial development. It is supporting low-cost biotechnology adoption programmes for socially backward communities. The programmes include vermi-composting, use of organic manure, silk-worm rearing, mushroom cultivation, etc. Some training cum demonstration programme are also being supported for them. Efforts are also on for gender mainstreaming.

The DBT has launched 11 projects for women in the areas of waste management, bio pesticides, bio fertilizers, floriculture, fish farming for poor women in the rural areas. At the Golden Jubilee Biotechnology Park for Women, industrial modules have been allotted to women entrepreneurs for setting up units in the above-mentioned areas.

This Park was inaugurated in 1998 near Chennai the capital of a southern Indian state Tamil Nadu. The Park is located in Siruseri, adjacent to Information Technology Park in an area of about 20 acres. It would have some central facilities for the entrepreneurs for technology resourcing, training and marketing. This is a first unique project of a joint effort by the central and state government in the realm of biotechnology.

ICAR and even non-ICAR research laboratories to work on crops like rice, cotton, brasica and brinzal. ICAR is also implementing a World Bank supported programme called National Agriculture Technology Programme (NATP) through which huge allocations have been made at different research laboratories for strengthening infrastructure for biotechnological research.

ICAR has also collaborated with DBT and Rockfeller Foundation to jointly launch a National Rice Biotechnology Network (NRBN) in 1988. This project helped in evolving a culture of collaboration among different institutions, which has led to publication of several international papers in established journals.

This network put together plant breeders and molecular biologists on the one hand and also provoked interests of private sector in R&D. For instance, Mahyco and Rockfeller Foundation worked together on identifying the relevant genes for saving Indian rice from brown plant hoppers. In this collaborative effort, interaction was not limited to private entities only. It is between private and public sector organisations also. Several universities also have tie ups with private sector organisations. For instance, Tamil Nadu Agricultural University (TNAU) attempted joint trials with Monsanto on weed resistance in soil. TNAU was also partner in monitoring the Bt Cotton field trials of Mahyco-Monsanto alliance. TNAU is also holding in exploratory talks with Rasi seeds (it is also in alliance with Monsanto) to conduct and monitor field trials of Bt crops. Similarly, rice research work at National Research Centre for Plant Biotechnology has attracted business interest of companies like Nath Seeds Ltd. and JK Agri-genetics.

## Human Resource Development and Training

The National Biotechnology Board had launched an integrated short-term training programme way back in 1984, to cope up with growing demand for highly trained manpower. In the first phase (1984-85), 5 universities were selected for initiating M.Sc./M.Tech programme in this multi-disciplinary area. Subsequently, in 1985-86 and 1986-87, the DBT has added 8 universities/ institutions for M.Sc/M.Tech/Post-doctoral teaching programmes.

Subsequently, DBT was entrusted with the responsibility of evolving curriculum for biotechnology courses and meet the demand for human resources in the field of biotechnology. In 1986-87 a model system of post-graduate/post-doctoral teaching in biotechnology in 7 universities/institutions was launched Some of the specialised M.Sc. courses in marine and agricultural biotechnology were launched in 1988-89 at 3 universities. In 1992-93, DBT supported a five year Integrated Programme in biochemical engineering and biotechnology in Indian Institute of Technology, Delhi and a post-doctoral programme at Indian Agricultural Research Institute, New Delhi.

Now, DBT is supporting 20 M.Sc. courses in general biotechnology, 4 in agricultural biotechnology, one each in medical and marine biotechnology while couple of diploma courses in molecular and biochemical technology 14. The total intake of students in the various postgraduate courses supported by the DBT in the country is around 550 per year.

As a part of restructuring of the post doctoral research and training programme, DBT has scraped the on going programme with different institutions and has given this responsibility to Indian Institute of Science (IISc), Bangalore. This is to ensure competitive attitude and quality output in the life sciences. It is being proposed that IISc would award up to 75 fellowships of two-year duration in different streams of biotechnology.

As part of a wider effort for capacity building in institutes of higher learning, full-fledged departments of biotechnology are being set up. The Indian Institute of Science, Bangalore, Indian Institute of Information Technology and Management, Gwalior; and select Regional Engineering Colleges are setting up departments of biotechnology. The All India Council for Technical Education (AICTE) has already approved B.Tech. programmes in biotechnology in eight engineering colleges and has since been advised to develop a model curriculum for undergraduate programmes.

All the new departments will have undergraduate, post-graduate and doctoral programmes. Special funding will be provided for the purpose in the Tenth Five Year Plan. Apart from expanding teaching of biotechnology at higher educational institutions a separate module on biotechnology would also be integrated with the school curriculum. The Department of Biotechnology of Government of India will provide the necessary outline of this module so that the National Council of Education Research and Training (NCERT) and the Boards of School Education would be accordingly advised.

Indian University Grants Commission has come out with a scheme to promote higher centres of learning at one place and assist them as much as possible. In this regard, Delhi based Jawaharlal Nehru University (JNU) has been identified by the UGC as centre for excellence in the areas of genomics, genetics and biotechnology. The University has received funds to the tune of ₹ 300 million and is planning to start a new integrated M.Sc./Ph.D programme in life sciences and made to upgrade equipment and library facilities.

The new integrated programme in life sciences will reduce the time taken by a scholar to complete his Ph.D by at least within two years. Now the scholar will not be required to undergo a separate M.Phil programme. JNU is aiming at 10 seats for the integrated course and another 20-25 seats in the School of Life Sciences.

The University has so far received 40 proposals for possible projects, which can be pursued by them in these fields in the future. Out of the funds that JNU has finally managed to get from the University Grants Commission on its selection as the University with potential for excellence, ₹ 100 million have been set aside for upgradation of facilities and equipments.

The rest of the ₹ 200 million would help University to subscribe to some of the 8,000 online journals, both in the filed of Science and social sciences. Besides this, JNU also announced recruitment of more researchers. The tie-ups with industry are also likely to grow. For instance, researchers at the Centre for Biotechnology (CBT) at Jawaharlal Nehru University in New Delhi have been working for four years on a recombinant anthrax vaccine and soon would start Phase I clinical trials in collaboration with Panacea Biotech Ltd. (New Delhi).

## FOREST BIOTECHNOLOGY

"The term biotechnology came into common usage in the 1980s, and its several definitions continue to change. Broadly defined, it is anything that combines biology and technology. Biotechnology has become more controversial as the level of the technology has increased." An excerpt from "Genetically Modified Forests from Stone Age to Modern Biotechnology" (Burdon et. al. 2006). This book is the most comprehensive review of forest biotechnology to date. Within the context of forest biotechnology we can refer to "modern" biotechnology as that which postdates the discovery of the structure of deoxyribonucleic acid, or DNA. Modern biotechnology is centered on the analysis and manipulation of DNA and the insertion of DNA fragments into organisms. Biotechnology has advanced to where we can research biological systems at the DNA level, and even sequence entire genomes. This feat was accomplished through years of research in plant and agricultural life sciences. While we can sequence a human's genome today, it is not as simple for some trees. For example, a representative conifer (pine tree) genome is many times larger than a human's (Keim, 2008); the Loblolly pine's genome is seven times larger. Such a feat has yet to be accomplished in whole, but the Pine Genome initiative is working towards this goal to improve future biofuel production from pine trees, improve forest health, and gain insight into plant evolution (Pine Genome Initiative, 2008). As we consider the spectrum of techniques that can be considered forest biotechnologies, it is useful to split out two general groups: breeding technologies, and genetic engineering technologies.

## Organogenesis

A type of asexual propagation that literally means 'organ genesis'. It is accomplished by growing a mass of cells in tissue culture that have the ability to produce shoots and grow into full trees.

## Somatic Embryogenesis

Cells in a tree other than the pollen and egg cells are called somatic cells. This technology-intensive technique is used to rapidly proliferate plant tissue via asexual propagation that mimics steps of the normal embryo development process. Single cells in tissue cultures from shoots, buds, roots, or leaves are induced to form complete embryos that can form individuals identical to the original plant where the cells were taken. Large numbers of a given clone can be produced in a short time with this technology.

## Breeding Technologies

Breeding technologies are described in this section to differentiate them from the scope of the Responsible Use: Forest Biotechnology Principles that will be discussed in more detail later in this chapter. The technologies are ordered from least to most advanced in terms of their reliance on advanced techniques for success. Be aware that one or more of these breeding technologies are used to produce genetically engineered trees.

## Conventional Breeding

Purposeful breeding of trees involves producing hybrids, progeny of parents from different genetic populations, to produce a desired phenotype. An example of this would be a cross between a tree that has exceptional growth and one that shows resistance to a fungus that attacks that particular species. This is an imprecise method of producing a tree with a specific characteristic because there is no control over additional genetic material being incorporated along with the desired phenotype.

## Asexual Propagation

Also referred to as clonal or vegetative propagation, is in contrast to sexual reproduction that requires fertilization via pollen. Sexual reproduction is how most plants reproduce. Asexual propagation produces genetically identical trees and does occur naturally by some tree species. For example, breaking off a branch of a willow tree and sticking it in the ground will produce roots from the branch. Over time a new willow that is genetically identical to the original tree will grow. Foresters have used this technology to their advantage as a breeding tool for centuries. Propagating a tree by grafting or with rooted cuttings allows the production of individuals that are genetically identical to the original and are therefore members of a clonal family. Tissue culture has become the preferred method of clonal production for some species of plants including several species of forest trees. This technology is also useful for moving clones across quarantine barriers since the cultures are free of insects and diseases and can easily be grown into mature plants.

### Marker Assisted Selection

When researchers have enough information about a tree's genome, they can use molecular markers to locate specific genes in a potential seedlings at an embryonic stage. These molecular markers are naturally in the tree's DNA, but they are not genes and they don't have a biological effect on the tree. These markers are easier to identify than specific genes themselves. By serving as constant landmarks in the genome, markers give breeders the tremendous advantages of speed and accuracy in their breeding program. With this technology they are able to determine if a tree, as an embryo, has the trait of interest instead of waiting tens of years for the tree to grow and mature.

## THE DIALOGUE OF FOREST BIOTECHNOLOGY

While it is impossible to rank-order every possible benefit and risk associated with a technology, we can reasonably expect society to make a distinction for trees that differ most from their native relatives based on the assumption that unfamiliar traits may pose more risk to humans or the environment. It will be society's concerns which will set the boundaries in the commercialization and stewardship of biotech trees.

In many ways, the source of society's concern involves speed. It has been argued, and will continue to be in some circles, that what humans are doing with genetic engineering is a faster version of the natural processes occurring all the time. Nothing is static. Species evolve over time. The 'natural' state of the earth's ecosystem is dynamic and always changing. For example, human beings are generally thankful for the mutations that have made us distinct from our ancestors. Together mutations and natural selection have created the myriad species on earth today, including the vast variety of trees in our forests. What we are able to do with biotechnology is twofold. We are able to speed up natural evolutionary processes that we can reasonably expect would happen over many millennia. An example of 'accelerated evolution' could be creating a tree that is resistant to a particular virus. We are also able to extend the abilities of trees in ways that we can reasonably expect would never happen naturally. An example of 'novel evolution' could be creating a tree that produces pharmaceutical products. Many of our drugs today are derived from tree products, such as tamoxifen (Geffen et. al. 2001), a drug originating from trees and used to fight breast cancer. In the future there may be trees that are engineered to produce products that are not naturally tree-based.

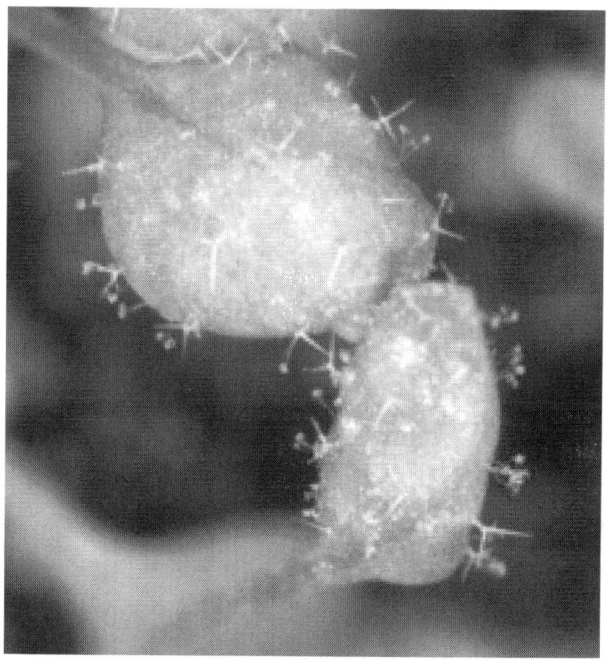

***Figure: Forest Biotechnology***

Biotech trees will change future forests. There are a number of ways that society can address the appropriate use of this technology: through laws and regulations, certification programs, and industrial pledges. Since society places a high value on the earth's forests and the ecosystem services they provide, we believe that social interests and the ecological need for science-based stewardship warrants more comprehensive options to address the use of forest biotechnology, and provide assurance that biotech trees can be used responsibly. To fill this gap in performance the Institute of Forest Biotechnology has launched an initiative that will create a set of principles for stewardship of these trees. This initiative focuses only on biotech trees; those depicted by the orange colors in the chart below. In very broad terms, increasing technical advances in biotechnology can potentially lead to greater deviations from native phenotypes. In other words, the more gene function is modified in a particular tree, the more likely that tree will have different characteristics from its native relatives. As the figure below shows, there is a higher potential to produce trees with characteristics that are different or altered from native phenotypes as we increase the amount of technology we apply.

However, more technology does not necessarily mean more genes are being modified. Many of the advances in biotechnology are to make gene modification more precise and thereby limit the extent of potential risks. This in an important distinction because a common belief is that applying more biotechnology equates to more significant changes in an organism's DNA. In reality, genetic engineering makes much more precise changes at the DNA level than traditional breeding. Even a considerable amount of genetic engineering would pale in comparison to the large DNA alterations that traditional breeding makes. It is the specific, targeted, and unique changes through biotechnology that make phenotype changes possible that differ from native ones. The graphic below is showing the potential changes from native phenotypes when more biotechnology is applied.

## RESPONSIBLE USE: FOREST BIOTECHNOLOGY PRINCIPLES

The following Principles are in recognition that responsibly used biotech trees have the potential to benefit society, the environment, economies, and cultures in ways that other trees cannot.

- Biotech trees will benefit people, the environment, or both
- Risks or benefit of biotech trees will be assessed
- Transparency is important-stakeholders will be engaged
- Social equity and indigenous rights are important and will be respected
- Biotech trees use must follow regulations of the appropriate country.

The Responsible Use initiative will help protect the future of our forests. In addition to the principles above, this initiative is grounded in basic truths about forests and biotechnology:

- Forests are important to people
- Biotechnology is a powerful tool
- Biotech trees have the potential for unique and diverse applications
- Biotech trees are being planted around the world
- Oversight of biotech trees is different around the world.

The framework of the Responsible Use initiative builds on these real-world realities with principles that provide a guiding sense of the requirements and obligations of using biotech trees responsibly. These principles are in recognition that responsibly used biotech trees have the potential to benefit society, the environment, economies, and cultures in ways that other trees cannot. Having a process that is transparent and inclusive of a wide range of perspectives is critical to determine what is 'responsible' and based on science. We also need practices that are not so prescriptive that a practitioner wanting to be a responsible user of forest biotechnology cannot afford to put these principles into action.

The IFB is managing this initiative by working with Sponsors, Forest Biotechnology Partners, experts from around the world, and any interested stakeholders.

Draft reports, meeting documents, and all other non-proprietary material relating to this initiative is available online at responsibleuse.org. There is also an online form where anyone interested in participating in the development of these principles can comment. These comments are available to the public for review; more importantly, they will be used by the Implementation Committee when developing the principles.

## Society Demands Sustainability

We need sustainably managed trees for communication, packaging, housing, food, and renewable energy. Currently the world does not have enough sustainably managed forests to fill all these needs. Instead we have illegal logging, land being converted from forests to sprawling housing developments, and an onslaught of invasive threats damaging the health of our forests. Forest biotechnology can be a powerful tool against these threats. While the debate of whether or not the use of biotech trees is sustainable carries on, the trees are being planted and used. As a non-profit organization based on science, dialogue, and stewardship, the IFB is committed to advancing the discussion of biotech trees and sustainability while at the same time developing a living set of principles that address the majority of current social concerns and scientific uncertainties. The future of our forests cannot be protected through inaction.

## Implementing the Principles

If we believe that society and the environment will benefit by having thoughtful experts create Responsible Use principles before any biotech trees are released into the natural environment, then we are already too late. As previously discussed, over a million Bt Populus have already been released into the environment in China, and as of printing, just one fruit bearing trees in the U.S. However, we still have the opportunity to significantly help to protect the future of forestry. There will be another wave of biotech trees in the next few years that will apply for commercial release (control level four)

Our goal is to have an initial set of Responsible Use: Forest Biotechnology Principles that practitioners can use to guide stewardship of biotech trees before they arrive on the market en masse: most likely in 2010 or 2011 at the latest. Involving a broad set of stakeholders from around the world, and testing the practices in real, on-the-ground situations is critical to developing a solid, broadly accepted, and highly streamlined set of practices. When they are available, practitioners will have a concise set of tools that have worksheets and guidance associated with each practice. There will also be a place to discuss with other practitioners in an open online forum at responsibleuse.org.

This initiative is designed to continue gathering input to refine the practices and add or remove them as necessary to keep abreast with forest biotechnology. Furthermore, the Responsible Use initiative strives to create a globally applicable set of principles. To do this we will coordinate with additional organizations and experts throughout the world.

# Bibliography


Andrew L. Winton and Kate Barber Winton : *A Handbook of Structure and Composition of Foods*, Agrobios Publication, Delhi, 2013.

Ashok K Chauhan and Ajit Varma:*Microbes : Health and Environment*, I K International, Delhi, 2007.

Aysha Aziz Faridi:*Dairy Microbiology*, Random Publications, Delhi, 2012.

Brandwein, P.F.: *Sourcebook for the Biological Sciences*, San Diego, Harcourt Brace JOvanovich, 1986.

Broach, J.R.: *The Molecular Biology of the Yeast*, Cold Spring Harbor, Cold Spring Harbor Laboratory, 1981.

Chrispeels, Maarten : *Plants, Genes and Crop Biotechnology*, Sudbury MA, Jones and Barlett Publishers, 2003.

Clark, J.M.: *Experimental Biochemistry*, New York, W.H. Freeman and Company, 1977.

Conko, Gregory: *The Frankenfood Myth: How Protest and Politics Threaten the Biotech Revolution*, New York, Praeger, 2004.

Cristobal Noe Aguilar; Juliana Morales Castro; Efren Delgado; Diana Jasso Cantu and Ashok Pandey: *Food Science and Food*

Dalip Kumar and Asmi Raza: *Agriculture and Food Security: Contemporary Issues*, Deep and Deep Publications, Delhi, 2011.

David Sadava: *Plants, Genes and Crop Biotechnology*, Sudbury MA, Jones and Barlett Publishers, 2003.

Davis, Thomas. 2000. *Sustaining the Forest, The People, and the Spirit.* Albany, NY: State University of New York Press.

Dixon, Dougal: *After Man-A Zoology of the Future*, New York, St. Martin's Press, 1981.

Doreen Virtue : *Constant Craving : What Your Food Cravings Mean And How To Overcome Them* , Hay House India, Delhi, 2011.

Edwards, C.A. *Sustainable Agricultural Systems.* 1990. Ankeny, Iowa: Soil and Water Conservation Society.

Franklin, Steven E. 2001. *Remote Sensing for Sustainable Forest Management.* Boca Raton, Fla.: Lewis.

Fransman M, Junne G, Roobeek A: *The Biotechnology Revolution?*, Oxford, Blackwell, 1995.

Fukuyama F.: *Our Posthuman Future: Consequences of the Biotechnology Revolution*, NewYork, Farrar, Straus and Giroux, 2002.

Fumento, Michael: *Bioevolution: How Biotechnology is Changing Our World*, San Francisco, Encounter Books, 2003.

Gadow, Klaus von and Pukkala, Timo. 2000. *Sustainable Forest Management*. Dordrecht; Boston: Kluwer Academic Publishers.

Gaskell G: *Biotechnology-the Making of a Global Controversy*, Cambridge, Cambridge University Press, 2002.

Goodsell, David S.: *Bionanotechnology: Lessons From Nature*, Hoboken, Wiley-Liss, 2004.

Harmeet Singh: *Dairy Farming*, APH Publication, Delhi, 2011.

Hatfield, Jerry L. and Karlen, Douglas L. 1994. *Sustainable Agriculture Systems*. Boca Raton: Lewis Publishers.

Madhulika Bhatnagar.: *Encyclopaedia of Catering Technology, Food Service and Hospitality Management, Vol. I to III*, Delhi, Anmol Publication, 2008.

Manju Malhi: *Easy Indian Cookbook: The Step-By-Step Guide to Deliciously Easy Indian Food at Home*, Viva Books, Delhi, 2008.

Manoj Negi: *Career in Dairy Farming*, Abhishek Publication, Delhi, 2010.

Piyush Sharma and Prem Ram : *A Textbook of Food and Beverage Management*, Dominant Publication, Delhi, 2013.

Tapeshwar Singh: *Resource Conservation and Food Security : An Indian Experience (2 Vols-Set)*, Concept Publication, New Delhi, 2004.

Tharakan : *A Young Hotelier's Guide to Food and Beverage*, Tata McGraw-Hill Publication, Delhi, 2005.

Udai Veer: *Elements of Food Science*, Anmol Publications, Delhi, 2007.

Urvashi Nandal : *A Handbook of Foods and Nutritional Biochemistry*, Agrobios Publication, Delhi, 2013.

Vishwambhar Prasad Sati: *Natural Resources Conservation and Food Security*, Bishen Singh Mahendra Pal Singh, Dehra Dun, 2012.

# Index